ENGLISH PLACE-NAME SOCIETY. VOLUME III.

THE PLACE-NAMES OF BEDFORDSHIRE & HUNTINGDONSHIRE

ENGLISH PLACE-NAME SOCIETY

The English Place-Name Society was founded in 1923 to carry out the Survey of English Place-Names, and to issue annual volumes to members who subscribe to thew work of the Society. The following county volumes have been published.

Volume I,   Part 1: *Introduction to the Survey of English Place-Names.*
             Part 2: *The Chief Elements Used in English Place-Names.*
             (Reprinted as a single volume).
Vol. II.   *The Place-Names of Buckinghamshire.*
Vol. III.   *The Place-Names of Bedfordshire and Huntingdonshire.*
Vol. IV.   *The Place-Names of Worcestershire.*
Vol. V.   *The Place-Names of the North Riding of Yorkshire.*
Vols. VI, VII.   *The Place-Names of Sussex, Parts 1 and 2.*
Vols. VIII, IX.   *The Place-Names of Devon, Parts 1 and 2.*
Vol. X.   *The Place-Names of Northamptonshire.*
Vol. XI.   *The Place-Names of Surrey.*
Vol. XII.   *The Place-Names of Essex.*
Vol. XIII.   *The Place-Names of Warwickshire.*
Vol. XIV.   *The Place-Names of the East Riding of Yorkshire & York.*
Vol. XV.   *The Place-Names of Hertfordshire.*
Vol. XVI.   *The Place-Names of Wiltshire.*
Vol. XVII.   *The Place-Names of Nottinghamshire.*
Vol. XVIII. *The Place-Names of Middlesex.*
Vol. XIX.   *The Place-Names of Cambridgeshire and the Isle of Ely.*
Vol. XX-XXII.   *The Place-Names of Cumberland Parts 1, 2, and 3.*
Vols. XXIII, XXIV.   *The Place-Names of Oxfordshire, Parts 1 and 2.*
Vols. XXV, XXVI.   *English Place-Name Elements, Parts 1 and 2.*
Vols. XXVII-XXIX.   *The Place-Names of Derbyshire, Parts 1, 2, and 3.*
Vols. XXX-XXXVII.   *The Place-Names of the West Riding of Yorkshire, Parts 1–8.*
Vols. XXXVIII-XLI.   *The Place-Names of Gloucestershire, Parts 1–4.*
Vols. XLII, XLIII.   *The Place-Names of Westmorland, Parts 1 and 2.*
Vols. XLIV-XLVIII.   *The Place-Names of Cheshire, Parts 1-4, and 5 (1:i).*
Vols. XLIX-LI.   *The Place-Names of Berkshire, Parts 1-3.*
Vols. LII, LIII.   *The Place-Names of Dorset, Parts 1 and 2.*
Vol. LIV.   *The Place-Names of Cheshire, Part 5 (1:ii).*
Vol. LV.   *The Place-Names of Staffordshire, Part 1.*
Vol. LVI/LVII.   *Cornish Place-Name Elements.*
Vol. LVIII.   *The Place-Names of Lincolnshire, Part 1.*
Vol. LIX/LX.   *The Place-Names of Dorset, Part 3.*
Vol. LXI.   *The Place-Names of Norfolk, Part 1.*
Vol. LXII/LXIII.   *The Place-Names of Shropshire, Part 1.*
Vol. LXIV/LXV.   *The Place-Names of Lincolnshire, Part 2.*
Vol. LXVI.   *The Place-Names of Lincolnshire, Part 3.*
Vol. LXVII/LXVIII/LXIX.   *The Place-Names of Rutland.*
Vol. LXX.   *The Place-Names of Shropshire, Part 2.*

In addition to subsequent Parts of the surveys of *Cheshire, Dorset, Lincolnshire, Norfolk* and *Shropshire,* Volumes for the following counties are in preparation: Co. *Durham, Hampshire, Herefordshire* and *Leicestershire.*

All communications regarding the Society should be addressed to:
The Hon. Secretary, English Place-Name Society, Department of English Studies,
University of Nottingham, Nottingham NG7 2RD.

THE SURVEY OF ENGLISH PLACE-NAMES
UNDERTAKEN WITH THE APPROVAL AND SUPPORT OF
THE BRITISH ACADEMY

THE PLACE-NAMES OF
BEDFORDSHIRE
AND
HUNTINGDONSHIRE

BY

A. MAWER and F. M. STENTON

NOTTINGHAM
ENGLISH PLACE-NAME SOCIETY
1996

Published by the English Place-Name Society
(Registered Charity No. 257891)
Department of English Studies
University of Nottingham NG7 2RD

ISBN: 0 904889 47 5

First published 1926
Reprinted 1969, 1996

Originally printed at the University Press, Cambridge
Reprinted in Great Britain
by Woolnough Bookbinding Ltd, Irthlingborough, Northamptonshire.

PREFACE

WHEN in the early days of 1925 the editors were faced with the difficult task of selecting a county for treatment in their series of volumes, they were peculiarly fortunate in receiving a very generous offer from Dr G. H. Fowler, the founder of the Bedfordshire Historical Record Society, who has done more than anyone else for the publication of the early records of that county. He undertook to provide in the requisite slip-form, either through himself or his helpers, the early forms and identifications of all place-names documented in the volumes of the Record Society and in Feudal Aids in addition. Further he undertook himself to supply in similar form the wealth of material to be found in certain documents which he. had himself transcribed, notably the Dunstable Cartulary and the Eyre Rolls for 1240 and 1247, both of which were at that time unpublished. With this solid foundation, consisting of all the fundamental material, the difficult task of completing a volume within a year at once became feasible. At the same time the editors realised that a volume on Bedfordshire alone would be considerably smaller than the average they had planned and unduly delay the progress of their great task. They therefore considered the possibility of associating some other county with it, and Huntingdonshire suggested itself as suitable in size and on historical and geographical grounds. The appointment of Miss E. G. Withycombe as secretary to the Director of the Survey made it possible, taken in combination with various other offers of help, to work through the remaining published documents for Beds and all for Hunts, and there only remained certain cardinal unpublished documents for the two counties alike. For this work there came the timely and generous help of the British Academy, which enabled us to secure the services of Miss Edith Scroggs, an old friend of the Society, in making excerpts from a few more Beds Eyre Rolls, several Hunts ones, the important Forest material for Hunts, many of the rich series of early Hunts Court Rolls, the Newnham Cartulary, packed with early Bedfordshire place-names, and to supplement Mr G. J. Turner's invaluable edition of the Huntingdonshire Feet of

Fines by a few field-names deliberately omitted from his edition.

Just when the editors were feeling that, though their task was a clear one and compassable within the time-limits imposed, the two counties might not provide very exciting fare in the nature of new discoveries there came a generous response from the Rev. Armitage Goodall (a name already well known to place-name students) to an invitation extended to him by the editors to let them publish certain most important discoveries he had made with reference to the early history of Huntingdonshire in his study of the Tribal Hidage.

When they decided to handle Bedfordshire and Huntingdonshire they were fully cognisant that they were entering upon ground already worked by a master in these studies. The justification for writing a volume after the two written by the late Professor Skeat is twofold. In the first place Professor Skeat dealt with (roughly) 200 names in Beds and 150 in Hunts as against (roughly) 400 and 250 respectively in the present volume, quite apart from the field-names which figure in this one alone. In the second place, and this is in some ways even more important, a great deal of material has now been printed which was not accessible in his days, and he, working single-handed, could not be expected to explore the wealth of early unpublished material which the resources of the Survey have opened up. This last point links itself interestingly, though not unexpectedly, with the fact that where the editors differ from Professor Skeat in their interpretation of a name it is, in most cases, solely as a result of this new material at their disposal. Without it their conclusions would as a rule have coincided with his, and the careful and close study of his volumes entailed in the preparation of their own has only brought home more clearly to them than ever how great and able a pioneer he was.

We now turn to the pleasantest part of our task, and that is once more to record our indebtedness to the many who, either of their own accord, or in response to the invitation of the editors, have given their help. At the head we must place the names of Dr Fowler and Professor Ekwall. The former has been unwearying in the clearing-up of all points of detail referred to him. To him we owe also the map of Bedfordshire which accompanies this volume. Professor Ekwall, as in all the previous work of the Society, has been fertile in suggestion, acute in criticism, and generous of time and labour.

Others we can but refer to in the alphabetical order of their names: Mr J. Hight Blundell for material for the history of Toddington drawn from his book on the history of that village and for other services.

The Rev. F. W. Breed for information with regard to Renhold and district.

Mr T. Candlin of Offord Cluny for detailed information with reference to the field-names of that parish.

Captain Cragg of Threckingham, Lincolnshire, for the use of the fragment of Anglo-Saxon MS recently discovered in his possession by Professor Stenton.

Mr O. G. S. Crawford for calling our attention to the important Fen Stanton charter not included in Birch or Kemble and first printed from the *Textus Roffensis* (c. 79) in *Proceedings of the Society of Antiquaries*, New Series.

Mr Bruce Dickins of the University of Edinburgh for many valuable criticisms and suggestions and for the use of his Leicestershire collections.

Mr J. Steele Elliott of Dowles Manor, Bewdley, for much-valued information as to the local pronunciation of names in Bedfordshire and for securing the help in this matter of Mr H. Howkins of Cople. To the latter also our cordial thanks are due.

Mr F. G. Emmison, Clerk of the Records, for transcribing field and other names from the Bedfordshire County Council muniments.

The Rev. Canon C. W. Foster for the use of his transcripts from the Lincoln Registers setting forth the institutions to benefices in the diocese and for collating certain forms.

Mr R. C. Gardner of Conington Hall, Cambridge, for compiling a Gazetteer of all the names on the 1-in. O.S. map of Huntingdonshire with many useful topographical and other notes and for answers to enquiries put to him.

Mr J. T. Gundry for local information with regard to Huntingdonshire.

Mr C. F. Hardy for identification of places recorded in Ogilby's *Itinerarium Anglicanum*.

The Rev. W. H. Harrison for help in transcribing forms from the *Red Book of Thorney* in the University Library at Cambridge and also for some forms from the St Neots Cartulary.

Mr J. N. Heathcote of Conington Castle, Huntingdonshire, for placing at our disposal names and identifications of places to be found in Court Rolls of Conington in his possession and for untiring readiness in endeavouring to help the editors to understand the always difficult topography of a parish which consists largely of drained fen-land.

Mrs Levett for help in transcribing forms from the volumes of the Bedfordshire Historical Record Society.

Miss K. M. Matthison for transcribing slips from various Huntingdonshire documents.

The Rev. Dr R. A. H. Murray, Rector of Broughton, and the farmers of his parish for information relating to the field-names of Broughton.

Mr H. E. Norris for the use of his unique collection of maps and books relating to Huntingdonshire. Unfortunately this collection only came to the notice of the editors as the volume was passing through the press and opportunities for making use of it were therefore limited by considerations of time and expense.

Mr Wm. Page, General Editor of the *Victoria County Histories* and, through him, his workers, for clearing up certain points of Huntingdonshire manorial history which offered difficulties and will be handled in full in the forthcoming *Victoria County History of Huntingdonshire*. The corresponding work on Bedfordshire was helpful at every turn in the work upon that county, and one learned only too fully to appreciate its value when one turned to Hunts for which there was no published volume.

The Rev. Edward Peake for much trouble in answering enquiries with reference to Bluntisham Parish.

The Rev. J. A. Ross for information with regard to early field-names in Holywell.

The President and Fellows of St John's College, Oxford, for the loan of the place-name collections of the late W. H. Stevenson, bequeathed by him to the College. So far it has only been possible to go through a part of them, but already in the present volume their usefulness begins to make itself apparent. In particular thanks are due to the College Librarian, Mr Austin Lane Poole, for his helpfulness in the whole matter.

Miss M. S. Serjeantson for transcribing the place-name material to be found in the Bedfordshire Charters in the Bodleian Library and for extracting from the transcripts of the Merton College Deeds made by the late W. H. Stevenson all relevant Bedfordshire material.

Mr C. C. Tebbutt of Bluntisham for valuable information with reference to Huntingdonshire topography and local pronunciations.

Miss Margaret White for help in transcribing forms from *Feudal Aids*.

Mrs Woods for transcription of forms from the *Red Book of Thorney*. It is with peculiar pleasure that we thus express our obligation to a daughter of the late Professor Skeat.

Professor R. E. Zachrisson for much time and trouble in helping to elucidate *cruces* presented to him. Our indebtedness to him is recorded in several places in the volume itself.

The officials of the Map-room at the British Museum for much-valued help in calling the attention of the editors to various early maps of the counties.

The County Organisers in general for their readiness in answering enquiries about comparative material, and to Mr J. E. B. Gover, Mr F. T. S. Houghton and Major J. de C. Laffan in particular for such offices in connexion with their collections for Devon, Worcestershire and Northamptonshire respectively.

The Director's secretary, Miss Withycombe, and Miss Scroggs, whose services were secured as noted above, the former for help given at every turn in the preparation of the volume, however monotonous and uninteresting the task may at times have been, and to Miss Scroggs for the results achieved during the four months that the Society was able to afford her services.

In response to certain happy suggestions made by Mr Bruce Dickins, Canon Foster, Professor Wyld and others, certain slight changes in the arrangement of the volume have been made as compared with that of the Buckinghamshire one.

Once again we have to record our thanks to Professors Ekwall and Tait for reading the proofs, with much profit to the volume. To their names have now to be added those of Mr Bruce Dickins of Edinburgh and Dr Fowler.

Finally we must record our obligations to the Cambridge Press for all their watchful care in the production of this as of all previous volumes of the Society's publications. It has done much to lighten our labours.

A. M.
F. M. S.

St Barnabas' Day, 1926.

CONTENTS

MAPS

INTRODUCTION

THE district covered by the present volume has a geographical rather than a historical unity. It is essentially the region traversed by the middle Ouse between its emergence from the uplands south of Yardley Chase and its subsidence into the level of the fens. To the south of this central region, the district includes the broken country between the Ridgmont-Ampthill escarpment and the Dunstable Downs. On the north it is bounded by the Nen between Elton and Peterborough. But the valleys of the Ouse and its tributaries determine its general character. It is a country in low relief, which can have presented few obstacles to the migration of peoples in early times.

In fact, the intermingling of different racial elements gives a peculiar character to the Anglo-Saxon archaeology of this region. The large burial ground at Kempston, for instance, yielded objects which prove the intermingling of Saxon and Anglian cultures in the Ouse valley before the end of the sixth century. Further to the east, the early burial grounds of Cambridgeshire have produced evidence pointing to a similar complexity of culture. Its historical significance is a very important, but most difficult, question. It is still uncertain whether this intermingling is due to the occupation of early Saxon sites by later Anglian settlers or to the eastward drift of Saxon objects into Anglian territory along the line of early travel known as the Icknield Way or even, as has recently been suggested, to the penetration of this region by Saxon settlers entering England by the Wash. The solution of this central problem will only begin to appear when the evidence derived from the historical, geographical, and archaeological aspects of the problem has been brought into coherent relationship. In the meantime, it may be suggested that the place-names of the region with which this book is concerned, and in particular the place-names of south Bedfordshire, deserve to be taken into account in any interpretation of the archaeological or historical evidence.

The recorded history of the region begins with the annal for 571 in the Anglo-Saxon Chronicle. It relates that a certain

Cuthwulf fought in that year with the Britons at *Bedcanford* and took four towns named *Lygeanburg, Aegelesburg, Baenesingtun,* and *Egonesham.* The appearance of this entry in a West Saxon set of annals, and the fact that the name Cuthwulf alliterates with the other personal-names used at this time in the West Saxon royal house, prove that the annal is describing a West Saxon conquest. There can be no doubt about the identification of the last three place-names mentioned in the annal with the modern Aylesbury, Bensington, and Eynsham. The forms given later in this volume are sufficient to establish the identification of the *Lygeanburg* of the Chronicle with the modern Limbury in south Bedfordshire. It is therefore evident that if the tradition preserved in the annal may be trusted, the West Saxons conquered at this time all the country north of the Chilterns between the Thames and the headwaters of the Lea. The traces of Saxon culture in the burial grounds of south Bedfordshire thus receive a simple historical explanation.

It is unfortunate that no suggestion can at present be offered bearing on the identification of the place named *Bedcanford,* where the battle of 571 was fought. The arguments against the long established identification with Bedford, an identification apparently first made by the historian Aethelweard in the tenth century, are set out later in the volume. The general course of West Saxon history in this age suggests that the battle was fought in the district to the east of the middle Thames, but nothing more definite can be said at present. The rejection of an ancient identification without the suggestion of something better is a thankless work, but it carries an important historical consequence in the present case. With the abandonment of the Bedford identification there disappears the only reason for believing that the West Saxons reached the Ouse valley in the sixth century. Limbury, the most northerly point of their recorded occupation, is separated from the Ouse by nearly twenty miles of varied country. The occupation of the district round the sources of the Lea does not in any way imply the occupation of the valley of the middle Ouse, and one serious complication is thus removed from the discussion of the remarkable objects found in the Kempston burial ground.

The general character of Bedfordshire place-names does not

suggest that the settlement of the county began before 571. The names Pillinge near Wootton and Kitchen (End) near Pulloxhill may well have arisen before the end of the sixth century. Knotting, in the north of the county, which almost certainly arose in this century, belongs in all probability to the Anglian settlement of uncertain but early date which produced Yelling and Gidding in Huntingdonshire, Kettering and Billing in Northamptonshire. Moreover, the detailed study of Bedford-shire place-names will probably create the impression that the settlement of this county was appreciably later than that of Buckinghamshire. In the latter county, Oving, Wing, and Halling, the obscure but certainly archaic Kimble, the traces of ancient personal nomenclature preserved in Mursley, Missenden, Haversham, and Mentmore, are only part of a body of evidence pointing to early settlement which is much more considerable than anything to be derived from the place-names of Bedfordshire. It may, indeed, be said that place-names which suggest early settlement seem to decrease in passing from Oxfordshire through Buckinghamshire into Bedfordshire, and to increase in passing from Bedfordshire through Cambridgeshire into Suffolk. It is, no doubt, true that the impression produced by a large number of names should not be accepted without reservation. Some of the Bedfordshire place-names which present no features suggesting high antiquity may still in fact have arisen at an early date. It can only be said that if the settlement of Bedfordshire began before the date assigned to the battle of *Bedcanford* by West Saxon tradition, the first settlers have left remarkably little trace upon the place-names of the county.

A late origin for this nomenclature is also suggested by the appearance of features which must have been very gradually developed. The large number of names ending in OE *hoh*, 'hill-spur,' sharply distinguishes the local nomenclature of Bedfordshire and Huntingdonshire from that of the adjacent counties. The prevalence of these names appears on a brief glance at the Ordnance map, though many ancient names originally of this type have now disappeared and many others have become disguised by the influence of common pronun-ciation upon spelling. The frequency of these names in the small area covered by this book is one of the most curious

B

features of the local nomenclature of the midlands. Its historical significance is that it gives a hint as to the length of the period over which the gradual settlement of this region extended. These names must come from a time when the invaders had been long enough in occupation of the country to have developed habits of their own in regard to local nomenclature. And one may be sure that local habits of this kind did not arise in less than a century from the beginnings of Anglo-Saxon settlement in this part of England.

The actual date of this settlement will always remain uncertain. The contribution which place-name study can make towards its establishment can only be indirect. But it may at least be said that the place-names of Bedfordshire, regarded as historical evidence, agree very well with the conclusions which are generally drawn from other forms of enquiry. The little that is known of the early history of the southern midlands suggests that two separate lines of invasion, the Saxon movement from the south-west and the Anglian movement from the north-east, met in what is now Bedfordshire. Place-name study cannot yield any definite evidence directly bearing upon this suggestion until it has become possible to identify particular types of name as characteristic of Saxon or Anglian settlement. When in the course of time the personal-names compounded in the place-names of the different parts of England have been subjected to a closer analysis than is at present possible, definite criteria of Saxon or Anglian settlement may begin to appear. But on the broader question whether the Anglo-Saxon occupation of Bedfordshire belongs to the earlier or later phase of the migration, the place-names of the county give valuable information. The extreme rarity of types of name which are known to be early, the undistinguished character of the local nomenclature, regarded as a whole, go far to prove that the county was not settled until the original energy of the invasion had spent itself. The place-names of Bedfordshire suggest the very gradual settlement of a tract of country rather than its rapid occupation by a large body of invaders. The inference naturally drawn from the Chronicle that the settlement of the south of the county was first made possible by Cuthwulf's victory of 571 agrees remarkably with the character of its local nomenclature.

It is highly probable that the settlement of Huntingdonshire was earlier than that of Bedfordshire. It is a small county, and woodland in the west and fen in the east must have limited very narrowly the land which was attractive to early settlers. Nevertheless, a very archaic element is apparent in its local nomenclature. It is remarkable that so small an area in the midlands should include four such ancient names as Gidding, Yelling, Lymage, and Wintringham. For historical purposes, the latter is one of the most useful names in the county. It occurs again as the name of a village in a fold of the Yorkshire Wolds, near the Derwent, and as the name of a village in Lindsey on the Humber. Both these sites must have been occupied at a very early date, and the Lincolnshire Winteringham may well represent a settlement founded by the earliest Anglian invaders. The occurrence of this name in Huntingdonshire therefore suggests a very early date for the settlement of this district, and confirms the probability suggested by its later history, that its original settlers were Angles.

Even more suggestive of early settlement is the name Earith, borne by a village at the point where the Ouse definitely enters the great tract of the Fens. The first element of this name can safely be identified with an OE word *ēar*, cognate with the ON *aurr*. Its exact meaning is uncertain, for in OE literature it only occurs as the name of one of the letters of the runic alphabet in a somewhat vague passage in the *Runic Poem*. Whether it means mud, or as Scandinavian parallels would suggest, gravel, it certainly denoted in the present case the first landing place available to travellers coming up the Ouse. The fact that the name of this landing place contains a word never found in strictly literary sources suggests that it arose at a very early date. The suggestion is confirmed by the fact that a duplicate of the name is found in Erith, Kent, upon the Thames, in a region which was probably settled in the fifth century. It would be unwise to assume that the Huntingdonshire name is as early as this, but it certainly agrees well with the other archaic names which distinguish the local nomenclature of this county.

In regard to one important matter the local nomenclature of Bedfordshire agrees with that of Huntingdonshire. The extreme rarity in either county of names which contain a Celtic element would of itself suggest that the Britons had disappeared from the Ouse valley some time before the end of the sixth century. Reasons are given below for believing that a Celtic element underlies the difficult name of Lattenbury Hill in Huntingdonshire, where the Ouse begins to enter the great level of the Fens, and the no less difficult name of Kempston in Bedfordshire, where a sixth century Anglo-Saxon settlement is certain on archaeological grounds. But these names stand almost alone. Place-names give no support to the theory, in itself improbable, of a long-continued British survival in the Fens. No name of any Romano-British road station has survived, in however mutilated a form, in the two counties. On the borders of Bedfordshire and Buckinghamshire Magiovintum has become Fenny Stratford, and the name of the Romano-British settlement which once existed at Sandy has vanished. In Huntingdonshire, Durolipons on the Ouse and Durobrivae on the Nen were renamed by Anglian settlers as *Godmundes ceaster* (Godmanchester) and *Deormundes ceaster*. Only the names of streams and rivers, and place-names such as Campton, Yelden, and Hail Weston, in which river-names are compounded, remain to prove the former existence of a British-speaking people in these shires.

Recent work has made it possible to recover something of the administrative geography of this region in the time before the counties of Bedford and Huntingdon were created. The counties themselves are of late origin. They are first mentioned by name in the year 1011, and each of them probably represents the district occupied by one of the Danish armies among which the southern Danelaw was divided before its conquest by Edward the Elder. The little that is known of the earlier divisions of this district comes from the enigmatical record now commonly known as the Tribal Hidage. Among many obscure names, three can now be definitely associated with the region with which this book is concerned. The three hundred families who formed the folk whose name appears in the genitive plural as *Gifla* were certainly inhabitants of the district of the Bedfordshire river Ivel. Mr Goodall has shown that the name of Hurstingstone Hundred preserves the folk-name which appears in the Tribal Hidage in corrupt forms such as *Here-finna, Hersinna, Herstinna.* He has also established a connexion between the difficult folk-name *Sweord ora*, which comes next in order in the Tribal Hidage, and Sword Point, the name of a low promontory on the edge of Whittlesey Mere. As only three hundred families are assigned to the latter folk, it may have been confined to the north of the modern Huntingdon-shire. But the *Hyrstingas* of Hurstingstone were twelve hundred families strong, and it is probable that their territory extended into the present Northamptonshire for some distance towards the west.

It is more remarkable that the Scandinavian invasion of the ninth century, which in many ways affected the history of these counties, has left little trace upon their local nomenclature. Huntingdonshire and Bedfordshire form with Cambridgeshire and Buckinghamshire a wide tract of country in which a Danish settlement, which must have been considerable, had little influence upon the forms of existing place-names or the creation of new ones. That all these counties formed part of the Danelaw, in the legal sense of this term, is certain. They are all definitely assigned to it in records of the early twelfth century, when the differences between the law of the Saxons, Mercians, and Danes were matters of immediate practical importance. Nevertheless, it would never have been inferred from the place-names of any of these counties that they had once undergone a Danish settlement sufficient to change the whole customary law which prevailed within them. In Bedfordshire, Clipstone certainly contains a Scandinavian personal-name, Stagsden and Renhold may possibly do so. Otherwise the Scandinavian element in the surviving place-names of this county is confined to a small group of local names of late origin found in Sharnbrook. In this connexion it should be observed that the ninth century boundary between Danes and English, which as defined by Alfred and Guthrum ran straight from the source of the Lea to Bedford, has no significance in regard to the local nomenclature of the county. Clipstone lies well on the English side of this boundary.

In Huntingdonshire, the Danish element is more apparent. It occurs in the names of two Hundreds, Normancross and

Toseland. Of these the latter is especially significant, for, unlike Clipstone, it consists of a Scandinavian word preceded by a Scandinavian personal name. It has the further interest that it may well preserve the name of the Danish earl Toglos who perished in the wars which ended in the conquest of this region by Edward the Elder. Scandinavian personal-names appear again in Keystone and Sibthorpe, the latter, like Toseland, is a name in which both elements are Scandinavian, and the Scandinavian *Kaupmaðr* occurs in Coppingford. The detailed study of the medieval field-names of the county certainly reveals many traces of Scandinavian influence which do not appear upon the modern map. In the perambulation of the king's demesne woods of Weybridge in 1227, the boundary passes through a place called *Kingstorth*[1], a name which undoubtedly contains the ON *storð*, 'wood,' common in forest-names in the northern Danelaw, but not otherwise found in Huntingdonshire. But it is probable that a minuter examination of Huntingdonshire field-names than is at present possible would only confirm the impression produced by the English character of surviving place-names, though it would doubtless produce interesting illustrations of the Anglo-Scandinavian personal nomenclature which had arisen in the wider Danelaw by the early Middle Ages.

The feudal names of Bedfordshire and Huntingdonshire are neither numerous nor distinctive. They are fewer, and on the whole of later origin, than the names of this class preserved in Buckinghamshire. Offord Cluny recalls the gift of an estate in this village made to the great abbey of Cluny by Arnulf de Hesdin before the date of the Domesday Survey. It is probable that the families of Longueville and Waterville were already established in the two villages which Domesday describes under the single name of *Ovretune*. Most of the remaining names belong to the thirteenth century rather than to the eleventh or twelfth, and the terminal *ing* of Westoning, which curiously disguises the name of William Inge, chief justice of the King's Bench in 1316–7, cannot have been annexed to the name of his manor of Weston before the reign of Edward II. It is unfortunate that the surname *Bussard*, which already in the thirteenth

[1] Rot. Litt. Claus. II, 209.

century was commonly appended to the name of the ancient royal manor of Leighton, still remains an unsolved problem.

Although the present volume deals with a region in which the great abbeys of the fen-land received early and large endowments, it has not derived a great store of pre-Conquest forms from their muniments. Most if not all of them were founded or refounded after the Danish wars of the tenth century, and obtained a considerable part of their property by purchase from men of moderate estate. Acquisitions made in this way were not confirmed by solemn diplomas with elaborate boundary clauses in English. Instruments of this kind were more appropriate to the conditions which prevailed in the south, where estates were more compact, and the royal power was more effective. In the Danelaw, the testimony of a hundred or group of hundreds and the proffer of sureties by the vendor were often considered sufficient security for the transfer of considerable estates, and a brief memorandum of the transaction made for the convenience of the new owner might well be the only written record of the proceedings. As such memoranda had little legal value, there was no need for them to be preserved with care when the new owner had become secured in possession by the passage of time, and there was even less reason why they should be transcribed into cartularies at a later date with the literal fidelity naturally demanded by formal, royal charters. It was therefore natural that when the monks of Ely and Ramsey wished to compile a record of the history of their respective houses, they were content to work the ancient memoranda which they possessed into a consecutive narrative without troubling to reproduce either the form of these documents or the Old English spellings of the place-names which occurred in them. The *Liber Eliensis* and the *Historia Ramesiensis* are more valuable as illustrations of social and legal history than is always recognised, but they have not the same importance as sources of Old English name forms.

For the rest, the materials used in the preparation of this volume are similar to those available for Buckinghamshire. The Domesday forms of place-names in Bedfordshire and Hunting-donshire contain few downright errors of the kind which complicated work in the latter county. Assize Rolls have proved invaluable once more. There is no unpublished cartulary relating to these counties so long or so important as that of Missenden, but the cartularies of Warden and St Neots have preserved important twelfth century documents in thirteenth century copies, and the cartulary of Newnham, written in the fifteenth century, has been especially useful for early forms of Bedfordshire field-names. As in Buckinghamshire, the interval between 1086 and 1200 has produced comparatively little material. But it would not be right to close this introduction without referring to the fortunate chance which brought to the notice of the editors the fragment of an original tenth century memorandum relating to the property acquired by bishop Aethelwold of Winchester in the district bordering on Whittlesey Mere. It is unhappily brief and badly mutilated, but through the courtesy of its possessor the editors are for once able to include in a volume forms derived from a pre-Conquest text which still remains unpublished.

NOTES ON THE DIALECT OF BEDFORDSHIRE AND HUNTINGDONSHIRE AS ILLUSTRATED BY THEIR PLACE-NAMES

OE æ appears normally as [æ] in modern names: *Halsey, Hatley, Hatch, Apsley, Gladley* in Beds. All these show ME forms in *a* except Gladley, but its etymology is doubtful (*infra* 124). Hunts has *Glatton* and *Haddon*, with one *e*-form for *Glatton*.

ME *a* has undergone characteristic lengthening to [aˑ] in the local pron. of such names as *Astwick, Astey, Aspley Guise, Salford*, and *Caldecote* in Northill.

OE *a* (Anglian) before *ld* is lengthened and then rounded to [ɔˑ], as in *Harrold* and *Wold* in Odell. So also *Old* Hurst and *Old* Weston and *Bromswold*. The one exception to this development is *Weald* in Eynesbury Hardwicke, where, from early times, we have a long run of *Weld*-forms going back to WS *weald*, side by side with the normal *wold*- ones. The explanation is possibly that in this royal forest area a non-local form established itself.

ME *a* for the *i*-mutation of OE *a* before a nasal is an East Saxon peculiarity with certain westward extensions (cf. Heuser, *Alt-London* 37, and Jordan, *ME Grammatik* 53). Heuser (40) noted certain unidentified examples in Bedfordshire. These may now be confirmed and extended. Cf. also Zachrisson in *Englische Studien*, LIX, 349. *Fancott* in Toddington in the south-west of the county and *Dane* End in Hyde in the south-east are clear examples. It has not prevailed in *Henlow* in the east of the county, but it has affected the early forms and it may explain some of the forms of *Hinwick* if that name contains OE *henn*. Examples of *dane* for *dene* have been noted in field-names in Biggleswade and Totternhoe, while *fan* for *fen* has been noted in Beeston, Ampthill, Flitton. There is no trace of this sound-development in Hunts, though there are very many examples of *fen*-names. *Tempsford* and *Kempston* with æ before *m* show long hesitation between *a*- and *e*-forms, while Bletsoe in which æ comes before *ch* always has *e*, except for one form.

Anglian æ for W S *ie* **before** *l* **followed by a consonant** is found in the early forms of *Wansford* and a lost *Wallpool* in Hunts. In the numerous examples of *wielle* (Angl. *wælle*) as the second element in a compound, the normal *well*-form is always found, except possibly in *Broadall's* District in Ramsey. These examples suggest that this dialectal feature, first fully studied by Ekwall in *Contributions to the History of the OE Dialects*, 40 ff., may have extended farther east than he makes it. It is difficult to explain the series of *Walm*-forms for Wansford, as against one *Welm*-form, on the ground of the influence of the verb *wallen*, 'boil,' as is done on p. 57 of Ekwall's paper.

In the **treatment of O E æ** it is difficult to reach any conclusions with regard to the development of WS *strǣt*, Angl. *strēt*. There are three examples on the eastern side of Bedfordshire. *Stratford* has only one early form, and that in *Strat-*, *Stratton* at first (with one exception, of somewhat remote provenance) has *Strat*-forms, then *Stret*-forms begin to appear but ultimately give place to *Strat-*. We must also note the interesting *Straittun* in a 12th cent. charter of definitely local provenance. *Streatley*, on the other hand, has æ in a Saxon charter and shows *a-* and *e*-forms from the earliest times. On the whole it would seem that the influence of ME *strete* as an independent word has tended to oust the *Strat*-forms which are the native ones. **East Saxon** *ā* **for** *ǣ* is clearly found in names containing *mǣd* as shown by the early forms of *Shortmead* (only one *mede*-form) and Wootton *Broadmead*. In *Bushmead* forms in *mede* and *made* are about equally common. In field-names *mede* and *made* are used indifferently. In *Whipsnade*, forms in *snede* and *snade* (from *snǣd*) are about equally common. *Wrest* has more *a-* than *e*-forms, but the latter ultimately prevail as they do also in *Slepe*.

There is a **tendency to unround O E** *o* **to M E** *a*, but the process is carried to completion only in *Clapham* and *Gadsey* Brook though various spellings bear witness to it in *Clophill*, *Stotfold*, *Toft*, *Totternhoe*, *Cogswell* in Beds. *Bolnhurst* and *Colworth* show *a* for *o* in early forms but the process is not carried so far as in the other names.

In *Chicksands* and *Sandy* we have a long struggle between **a- and o-forms before** *nd* and *o*-forms have prevailed in *Stondon*. So also *le sondes* in Elton (1300).

OE *u* appears as *e* before point-consonants in *Denel End*, *Dedmansey* and *Seddington* in Beds and in the *Deddington*-forms which preceded the present *Diddington* in Hunts. This is not, strictly speaking, a dialectal feature, as can be seen in the further examples quoted under these names in the text. It is noted here as the phenomenon does not seem hitherto to have been observed. The process was probably hastened by the presence of initial *d* or *s* before the vowel.

OE *y* normally appears as *i* in both counties alike. The only exceptions are *Luton* and possibly *Runley* in Luton (the etymology is obscure), where after some hesitation the *u*-forms prevail, and *Pedley* in Clophill, for which unfortunately we have only one early form and that a *u*-one and then the modern form in *e*. *Husborne* Crawley is difficult. The only OE form has *y*, all the ME forms show *u* with but two exceptions. Some process of analogy or folk-etymology would seem to have been at work here.

Taking the rest of the names in Beds and working up from the south *Bidwell* in Houghton Regis in early forms always has *u* except for one *e*. *i* is a late development. *Limbury* shows one or two *u*-forms. *Tilsworth* shows *i*- and *u*-forms, very rarely *e*-forms. *Shillington* on the east side of the county has *e*-, *u*- and *i*-forms equally commonly at first, but the *u*-forms disappear in the middle of the 14th cent. *Lidlington* has *u*- and *i*-forms, *e*-forms only very rarely. No *u*-forms after 1358. The influence of the independent word *little* has probably affected the history of this name. *Millbrook* on the west of the county agrees with *Millow* on the east in showing early forms in *e* and *u*. Those in *i* are comparatively late. The common word *mill* has certainly had its influence here. The lost *Kinwick* in Sandy shows *e* and *i*, never *u*, and the same is true of *Stevington* on the west side of the county, but with regard to this last name Professor Zachrisson points out that even in Stewkley (Bk) we never get *u*-forms, still less in Stukeley (Hu). *Biddenham* shows occasional *u*- and *e*-forms, usually *i*.

Passing into Huntingdonshire Dillington shows an *e* in DB, one *u*-form and then *e*- and *i*-forms which exist side by side till the 16th cent. *Kimbolton* has only *i*- and *e*-forms. The latter have not been noted after 1260, but here as in *Kinwick* the natural tendency would be to raise *e* to *i* before *n*. *Brington* has *e* in DB and then always *i*, except for a single *u*-form. The *Giddings* show regular *e* till the middle of the 13th cent. when *i* first appears and, by the middle of the 16th cent., ousts *e*. *Needingworth* shows hesitation between *e*- and *i*-forms. The former, with lengthening of the vowel, ultimately prevailed. This is on the eastern edge of the county. *Pidley* shows *i*-, *u*- and occasional *e*-forms. *Stibbington* in the north-west of the county has *e*- and *i*-forms, never *u*.

Bury, hurst, hill and *herne* can only be used satisfactorily in considering questions of dialect when they bear the stress. *Bury* (Hu) regularly has *i* in all its early forms. *Hurstingstone* has *e*-, *u*- and *i*-forms to begin with but the *e*-forms soon disappear. *Herne* in Ramsey has only *e*- and *i*-forms. The regular form for *hyll* in Beds is *hull* in ME, though this has now always given place to *hill*. Only once, however, do we get this in the 13th cent. In all other cases it is of much later date. *Hilton* in Hunts, on the other hand, shows only one *u*-form.

O E *eo* **has become** *i* and *e* respectively in *Chibley* in Shillington and in *Tebworth*. The former shows occasional forms in *u*. Possibly also *Hulsterdene* in Northill from OE *heolstor*.

O E *ēo* **has given certain forms in** *iu* **and** *u* **in the early history** of *Flitt, Flitton* and *Flitwick, Girtford* in Beds and once in *Fletton* (Hu). Note also *Dupedel* in Sandy (12th). Much more common are forms in *e*.

In the **treatment of initial** *c* there is considerable fluctuation. The numerous *Caldecotes* in both counties always show initial *c* in all their forms. Compounds of OE *ceald* (Anglian *cald*) with *wielle* yield *Chadwell* End in Pertenhall, but *Cauldwell* in Bedford. Among minor names we have *Chaldewelle* in Houghton Regis (1225), in Beeston (13th), *Chadewelle* in Stotfold (13th), *Cheldewell* in Stilton (1241) and *Chaldewelle* and *Coldewelle* alike in Odell in 1367, *Caudewelle* in Bythorn (1248). Other compounds with Anglian *cald* are found in *Caldelane* in Ridgmont (13th), *Caldemowe* in Warboys (1251). OE *cealf* (Angl. *calf*) yields *Chawston* in Beds together with minor names *Chalfhamstede* in Eaton Socon (1359), *Chaluescrofte* (13th) in Roxton, but we have *Calfhale* in Sharnbrook (13th), *Calvescroft* in Clophill (1273) in Beds and *Calpher* in Hunts, together with *Calurecroft* in Ripton (1252). Of compounds with *cealc* we have Chalgrave (Beds) and *Chalpet* in Ellington (1279). *ceaster* appears uniformly as *Chester* in our counties. Noteworthy is the fact that Chesterton (Hu) is only two miles from Castor (Nth), just across the Nen[1].

The **inflexional** *n* of the suffix **of the weak form of the adj.** has survived in *Newnham*, *Sharpenhoe* and *Whitnoe* in Beds and would probably have remained in *Langnoe* if that name had survived. Water *Newton* in Hunts does not show it In fieldnames we have *Druyhenhull* (1461) and *Drynehull* (1519) in Beeston, *Dryenholm* in Caldecote in 1331

[1] Stevenson MSS.

ABBREVIATIONS

Abbr	*Placitorum Abbreviatio*, 1811.
AC	*Ancient Charters* (Pipe Roll Soc.), 1888.
AD	*Catalogue of Ancient Deeds*. (In progress.)
AD	Ancient Deeds (unpublished) in PRO.
AN	Anglo-Norman.
ASC	*Anglo-Saxon Chronicle*.
Ashby	Cartulary of Ashby Canons Priory (taken by courtesy of Dr G. H. Fowler from MS in possession of Colonel Orlebar of Hinwick[1]).
Ass	Assize (or Eyre) Rolls for 1202, 1227, 1240 (BHRS ed.).
Ass	Assize (or Eyre) Rolls (unpublished) for Beds for 1247, 1276, 1287, 1307, for Hunts for 1227, 1260, 1286, 1314.
Award	Enclosure Award.
B	Bryant, *Map of the County of Bedford*, 1826.
BCS	Birch, *Cartularium Saxonicum*, 3 vols., 1885–93.
Berks	Berkshire.
BHRS	*Bedfordshire Historical Records Society*. (In progress.)
Bk	Buckinghamshire.
BM	*Index to the Charters and Rolls in the British Museum*, 2 vols., 1900–12.
Bodl	Bodleian Charters. (The abbreviation for the county is added after *Bodl*.)
Bowen	Emmanuel Bowen, *The Large English Atlas*, c. 1750.
Bract	*Bracton's Note-book*, 1887.
BT	Bosworth-Toller, *Anglo-Saxon Dictionary*, 1882–98.
C	Cambridgeshire.
Cai	*Admissions to Gonville and Caius College*, ed. Venn, 1887.
Cary	Cary, *English Atlas*, 1787.
Ch	Cheshire
Ch	*Calendar of Charter Rolls*. (In progress.)
ChancP	*Chancery Proceedings in the reign of Elizabeth*, 3 vols., 1827–32.
Chanc Roll	Variant readings from Chancellor's copy of Pipe Rolls as noted in Pipe Roll Society's volumes.
ChR	*Calendarium Rotulorum Chartarum*, 1803.
Chron Rams	*Chronicon Abbatiae Rameseiensis* (Rolls Series), 1886.
Chron St Alb	*Chronica monasterii S. Albani* (Rolls Series), 1853–76.
Cl	*Calendar of Close Rolls*. (In progress.)
ClR	*Rotuli Litterarum Clausarum*, 2 vols., 1833–44.
Co	Cornwall.
Cor Reg	*Coram Rege Rolls*.
Coroner	*Coroners' Rolls* (Selden Soc. vol. 9), 1896.
Cott	Cotton Charters in the British Museum.
Cragg	MS of Captain W. A. Cragg, Threckingham, Lincs.
Crawf	*Crawford Charters*, ed. Napier and Stevenson, 1895.
CS	Clerical Subsidies (BHRS i, 27–61).
Ct	Court Rolls (unpublished) in BM, PRO and in private possession.
Cu	Cumberland.
Cur	*Curia Regis Rolls*. (In progress.)

[1] Now in Brit. Mus. (Egerton 3033).

Cur(P)	*Three Rolls of the King's Court* (Pipe Roll Soc.), 1891. *A Roll of the King's Court* (ib.), 1900.
D	Devon.
D	*Letters and State Papers Domestic*, 12 vols., 1856–72.
Dan	Danish.
D and C Linc	Dean and Chapter of Lincoln MSS.
Darton	Wm. Darton, *English Atlas*, 1822.
DB	*Domesday Book*
Db	Derbyshire.
Do	Dorset.
Du	Durham.
Dugd	Dugdale, *Monasticon*, 6 vols. in 8, 1817–30.
Dunst	*Annales prioratus de Dunstaplia* (Rolls Series).
Dunst	Dunstable Cartulary (Harl. 1885).
EDD	*English Dialect Dictionary*.
EDG	Wright, *English Dialect Grammar*, 1905.
EME	Early Middle English.
EPN	*Chief Elements in English Place-names*, 1923.
Ess	Essex.
FA	*Feudal Aids*, 6 vols., 1899–1920.
Fees	*Book of Fees*, 2 vols., 1922–3.
FF	Feet of Fines for Beds from BHRS vi, for Hunts, as ed. by G. J. Turner for the Cambridge Antiquarian Society.
FF	Unpublished Feet of Fines.
Fine	*Calendar of Fine Rolls*. (In progress.)
For	*Select Pleas of the Forest* (Selden Soc., vol. 13), 1901.
For	Pleas of the Forest (unpublished).
Forssner	*Continental Germanic Personal Names in England*, 1914.
Förstemann	*Altdeutsches Namenbuch, Personennamen* (PN), *Orts-namen* (ON), 2 vols. in 3, 1901–16.
Fr	French
G	Greenwood, *Map of the County of Huntingdon*, 1831.
Gaimar	*Lestorie des Engles* (Rolls Series), 2 vols., 1889–90.
Germ	German.
Gerv	*Gervasius Cantuariensis* (Rolls Series), 2 vols., 1879–80.
Gest St Alb	*Gesta Abbatum Monasterii S. Albani* (Rolls Series), 3 vols., 1867–9.
Gilb	*Charters relating to Gilbertine Houses* (Lincoln Rec. Soc.).
Gl	Gloucestershire.
Ha	Hampshire.
HarlCh	Harleian Charters in the BM.
Harrold	Harrold Cartulary (Lansdowne MSS 391).
He	Herefordshire.
Herts	Hertfordshire.
HH	*Henrici Huntendunensis Historia Anglorum* (Rolls Series), 1879.
Higden	*Polychronicon* (Rolls Series), 9 vols., 1865–86.
Hist El	*Historia Eliensis*, ed. Stewart, 1848.
HMC	*Historical MSS Commission Reports*. (In progress.)
Inq aqd	*Inquisitiones ad quod damnum*, 1803.
Inq El	*Inquisitio Eliensis*, ed. Hamilton, 1876.
Inv	*Edwardian Inventories for Bedfordshire* (Alcuin Club), 1905.
Ipm	*Calendar of Inquisitions post mortem*. (In progress.)
IpmR	*Inquisitiones post mortem* (Record Commission), 4 vols., 1806–28.

IPN	*Introduction to the Survey of English Place-names*, 1923.
J	Jefferys, *Map of the County of Huntingdon*, 1768.
Jury	Jury Lists (BHRS iv, 61–255).
K	Kent.
KCD	Kemble, *Codex Diplomaticus Aevi Saxonici*, 6 vols., 1839–48.
KF	Knights Fees for Beds, 1303 (PRO).
KS	Knight Service (BHRS ii, 245–63).
L	Lincolnshire.
La	Lancashire
Lei	Leicestershire.
LGerm	Low German
Linc	Lincoln Episcopal Registers.
LP	*Letters and Papers Foreign and Domestic.* (In progress.)
LS	*Subsidy collected in the Diocese of Lincoln in* 1526 (Oxf. Hist. Soc.), 1909.
	Lay subsidy (1220), (BHRS ii, 225–38).
LVD	*Liber Vitae Dunelmensis* (Surtees Soc.), 1841.
Lysons	*Magna Britannia*, vol. i, 1806.
ME	Middle English.
Merton	Merton College Deeds, transcribed by W. H. Stevenson.
Misc	*Calendar of Inquisition Miscellaneous*, 2 vols., 1916.
MLR	*Modern Language Review.*
Mon	Monmouthshire.
Nb	Northumberland.
NED	*New English Dictionary.*
Newn	Newnham Cartulary (Harl. 3656).
Nf	Norfolk.
NI	*Nonarum Inquisitiones*, 1807.
NLC	*Newington Longueville Charters* (Oxfordshire Rec. Soc.). 1922.
NoB	*Namn och Bygd.*
Norw	Norwegian.
NQ	*Bedfordshire Notes and Queries*, 3 vols., 1886–93.
Nt	Nottinghamshire.
Nth	Northamptonshire.
O	Early ed. of O.S. maps.
O	Oxfordshire.
ODan	Old Danish.
OE	Old English.
OFr	Old French.
Ogilby	Ogilby, *Itinerarium Anglicanum*, 1675.
OHG	Old High German.
ON	Old Norse.
Ord	*Orderici Vitalis Hist. Eccl.*, 5 vols., 1838–55.
Orig	*Originalia Rolls*, 2 vols. 1805–10.
O.S.	Ordnance Survey.
OSwed	Old Swedish.
(p)	Place-name form derived from pers. name.
P	*Pipe Rolls* (Pipe Roll Soc. eds. in progress and, for Beds, BHRS, vii).
P	Pipe Rolls (unpublished).
Pat	*Calendar of Patent Rolls.* (In progress.)
p.n.	place-name.
PRO	Public Record Office.
QW	*Placita de quo Warranto*, 1818.

c

R	Rutland.
Rams	*Cartularium Monasterii de Rameseia* (Rolls Series), 3 vols., 1884.
RBE	*Red Book of the Exchequer*, 3 vols., 1896.
Redin	*Uncompounded Personal Names in Old English*, 1915.
Reg Dun	*Reginaldi Monachi Dunelmensis libellus* (Surtees Soc.), 1835.
RG	*Metrical Chronicle of Robert of Gloucester* (Rolls Series), 2 vols., 1887.
RH	*Rotuli Hundredorum*, 2 vols., 1812–18.
Ritter	*Vermischte Beiträge zur Englischen Sprachgeschichte*, 1922.
Rot Dom	*Rotuli de Dominabus* (Pipe Roll Soc.), 1913.
RW	Roger of Wendover in *Flores Historiarum* (Rolls Series), 3 vols., 1890.
s.a.	sub anno.
Sa	Shropshire.
St Neot	Cartulary of St Neots (Cotton Faust A iv).
St P	*The Domesday of St Paul's* (Camden Soc.), 1858.
Saxton	Saxton, *Map of the County of Huntingdon*, 1576.
Scand	Scandinavian.
Searle	*Onomasticon Anglo-Saxonicum*, 1897.
Selden	Selden Soc. Publications. (In progress.)
Sf	Suffolk.
Smith	Smith, *English Atlas*, 1804.
So	Somerset.
Speed	*Map of the County of Huntingdon*, 1610.
Sr	Surrey.
SR	Exchequer Subsidies for Beds (1297), for Hunts (1303, 1327), unpublished.
St	Staffordshire.
Stevenson MSS	Collections bequeathed to St John's College, Oxford, by W. H. Stevenson.
Strip	Strip-maps of Beds parishes (or copies of same), *pen.* Dr G. H. Fowler.
Swed	Swedish.
Sx	Sussex.
Tax	*Taxatio Ecclesiastica*, 1802.
Templars	Feodary of the Templars (Exch. K. R. Misc. Books, Series 1, vol. ii).
Terr	Terrier.
Thorney	*Cartularium et Registrum Coenobii Thorneyensis* (Camb. Univ. Lib.).
Thorpe	*Diplomatarium Anglicum Aevi Saxonici*, 1865.
TRE	Tempore Regis Edwardi.
TRW	Tempore Regis Willelmi.
VCH	*Victoria County Histories of England*. (When no further ref. is added the ref. is to the *History of Bedfordshire*, 3 vols. and Index, 1904–14.)
VCH	Material for the forthcoming VCH for Hunts (*ex inf.* the General Editor).
VE	*Valor Ecclesiasticus*, 6 vols., 1810–34.
W	Wiltshire.
Wa	Warwickshire.
Warden	Warden Cartulary (John Rylands Library).

We	Westmoreland.
Wells L	*Lib. Antiq. temp Hugonis Wells* (ed. Gibbons), 1888.
Whet	*Registrum abbatiae Joh. Whethamstede* (Rolls Ser.), 1872.
Winton	Winchester Episcopal Registers.
WMP	*Wm. Malmesburiensis Gesta Pontificum* (Rolls Series), 1870.
Wt	Isle of Wight.
Y	Yorkshire.

Reference is made to the various county place-name books already published (*v.* summary bibliography in *Chief Elements of English Place-names*) by using the abbreviation PN followed by the recognised abbreviation for the county, *e.g.* PN Gl for Baddeley's *Place-rames of Gloucestershire.*

PHONETIC SYMBOLS USED IN TRANSCRIPTION
OF PRONUNCIATION OF PLACE-NAMES

p	*p*ay	z	*z*one	r	*r*un	e	*r*ed
b	*b*ay	ʃ	*sh*one	l	*l*and	ei	fl*ay*
t	*t*ea	ʒ	a*z*ure	tʃ	*ch*urch	ɛː	*there*
d	*d*ay	θ	*th*in	dʒ	*j*udge	i	p*i*t
k	*k*ey	ð	*th*en	ɑː	*father*	iː	f*ee*l
g	*g*o	j	*y*ou	ɑu	c*ow*	ou	l*ow*
ᴧ	*wh*en	χ	lo*ch*	ai	fl*y*	u	g*oo*d
w	*w*in	h	*h*is	æ	c*a*b	uː	r*u*le
f	*f*oe	m	*m*an	ɔ	p*o*t	ᴧ	m*u*ch
v	*v*ote	n	*n*o	ɔː	s*aw*	ə	*o*ver
s	*s*ay	ŋ	si*ng*	oi	*oi*l	əː	b*i*rd

Examples:

Harwich [hæridʒ], Shrewsbury [ʃrouzbəri, ʃruːzbəri],
Beaulieu (bjuːli).

NOTES

(i) The names are arranged topographically according to the Hundreds working approximately from the north-west to the south-east of the county. Within each Hundred the parishes are dealt with in alphabetical order and within each parish the place-names are arranged similarly. The only exceptions to this rule are that river- and road-names are taken at the beginning as also the place-names in Bedford town itself, while Kensworth, once in Herts, is taken at the very end.

(ii) After the name of every parish will be found the reference to the sheet and square of the 1-in. O.S. map (Popular Edition) on which it may be found. Thus, **Alwalton** 74 A 10.

(iii) Where a place-name is only found on the 6-in. O.S. map this is indicated by putting (6″) after it in brackets, e.g. **The Green** (6″).

(iv) The local pronunciation of the place-name is given, wherever it is of interest, in phonetic script within squared brackets, e.g. **Astey** [aˑsti]. Where the old spellings indicate a local pronunciation which cannot now be traced but which must have prevailed at an earlier stage in the history of the name, that pronunciation is given in phonetic script as in the case of the other names, but it is preceded by the word *olim*, e.g. **Oakley**, *olim* [ɔkli].

(v) In explaining the various place-names summary reference is made to the detailed account of such elements as are found in it and are explained in the *Chief Elements of English Place-names* by printing those elements in Clarendon type, e.g. **Milton**, *v.* **middel, tun**.

(vi) In the case of all forms for which reference has been made to unprinted authorities, that fact is indicated by printing the reference to the authority in italic instead of ordinary type, e.g. *Ass* 1287 denotes a form derived from MS authority in contrast to FA 1284 which denotes one taken from a printed text.

(vii) Where two dates are given, e.g. 979 (12th), the first is the date at which the document purports to have been composed, the second is that of the date of the copy which has come down to us.

(viii) Where a letter in an early place-name form is placed within brackets, forms with and without that letter are found, e.g. *Sto(c)kton* means that forms *Stockton* and *Stokton* are alike found.

(ix) All OE words are quoted in their West-Saxon form unless otherwise stated.

ADDENDA ET CORRIGENDA
TO VOLUMES I & II

In this list the editors have confined themselves for the most part to points in which the volumes as issued contain statements which stand definitely in need of correction. Conditions of space have compelled them to omit reference to a good deal of comparative and other material which has since come to hand or been brought to their notice. In making these additions and corrections they are indebted to a number of friendly critics, including Mr Arthur Bonner, Mr Bruce Dickins, Professor E. V. Gordon, Dr G. B. Grundy, Mr F. T. S. Houghton, Mr O. Schram, and Professor Zachrisson.

VOL. I, PART I[1]

p. 101, l. 16, delete '*Sister*,'
p. 119, ll. 21–24, delete ', and in this connection...Old English times'
p. 122, ll. 4–6, 15–17, delete the references to Peperharow and Blo' Norton.
p. 131, l. 12, insert 'Hugh de Gunville, 1233'
p. 131, l. 20 from bottom, read 'Shillyngeston John Eskelling, 1226 FF' and for footnote 5 read '1444 (Ancient Indictments).'
p. 132, l. 22, add
'Tuz Seinzton   Lucas de Tuz Seinz, 1242   Towsington (in Exminster)'
p. 146, l. 12 from bottom, for 'the only example' read 'almost the only certain example'
p. 149, l. 20 from bottom, read 'Oxenhoath'
p. 152, s.n. *tunsteall*, I. Ancient. omit 'Bosworth-Toller...occurs there.' and re-number the sub-paragraphs.

VOL. I, PART II

p. viii, l. 17 from bottom, read 'Cambridge, 1901.'
p. 3, l. 6, for 'glossed as' read 'glossing'
p. 4, l. 9 from bottom, insert 'Little' before 'Barford'
p. 14, l. 18, add Hanscombe (Beds)
p. 17, l. 12 from bottom, for 'Hockley' read '*Hocheleia*, the DB form of Hockliffe'
p. 19, l. 6, cran. Mr Bruce Dickins and Mr W. L. Orgill both suggest that in some cases the word *crane* may have its common dialectal significance of 'heron.' The crane was never common except in marshy places.
p. 25, l. 20 from bottom, read 'sandy or gravelly bank, or beach, or spit of land' in place of 'sandbank'
p. 26, l. 20, for 'Farcett (Ha)' read 'Farcet (Hu)'
p. 26, l. 21. From the Stevenson MSS we may quote the apt passage

---

[1] One or two of the corrections to Part I have already been made in a second imprint of this volume.

found in *Pauli Diaconi Historia Langobardorum* (i, c. 26), '*Egressi quoque Langobardi de Rugiland habitaverunt in campis patentibus, qui sermone barbarico* "feld" *appellantur.*'

p. 27, l. 2 from bottom, for 'Harvington' read 'Harvington-on-Avon'
p. 28, l. 3, for 'fox-hole, ME' read 'fox-hol, OE'
p. 30, ll. 8–10, read 'The substantival use of this word to denote a "grassy spot," "village green" belongs to post-Conquest times.'
ib. l. 5 from bottom, for 'Licky' read 'Lickey'
p. 33, l. 20 from bottom, read 'found as *honger, hunger* in ME'
ib. l. 4 from bottom, read 'Harrock (La)'
p. 34, l. 12 from bottom, healh. Mr Bruce Dickins calls attention to the passage in Simeon of Durham (ii. 53) in which we have the phrase *Hearrahalch, quod interpretari potest locus Dominorum* which suggests that we ought not perhaps always to press for too precise a meaning for *healh.*
p. 35, l. 19, read 'Harrowden (Nth)'
p. 38, l. 3 from bottom, delete 'It seems to be unknown in England South of the Thames.'
p. 40, l. 15 from bottom, read 'Monyhull (Wo)'
p. 46, l. 7, add here the reference to Reginald of Durham's *nemus paci donatum* given in *PN Beds Hunts.*
p. 46, l. 14, for 'grass-land' read 'hay-land'
ib. l. 19, insert 'wielle' after 'weg'
p. 47, l. 10, read 'Monyhull (Wo)'
p. 49, l. 12, read 'Lapal (Wo)'
p. 50, l. 10. The word is on record in OE itself, v. *geryd* in BT supplement.
p. 53, l. 9 from bottom, for 'slá(h)' read 'sla(h)'
p. 54, delete ll. 1 and 2.
p. 56, l. 3, delete 'Oxted (Sr),' It belongs properly to p. 55, l. 6 from bottom. Insert there 'Oxted (Sr) from ac'
p. 56, l. 15 from bottom, stoc. Mr Bruce Dickins calls attention to the gloss upon Woodstock (O) in Simeon of Durham (ii. 267), '*Wdestoc, quod Latine dicitur Silvarum Locus.*'
p. 58, l. 5 from bottom, read 'Ede Way (Beds)'
p. 61, l. 3, treo. Mr Bruce Dickins suggests that in some cases the sense may be 'cross,' cf. *Croes Oswallt*, the Welsh name of Oswestry.
p. 61, l. 12, delete 'Trundle Mere (C)'
p. 63, l. 6, read '*sceap-wæsce*'

VOL. II

p. 2, l. 18 from bottom, read '*Ciltern-*'
p. 17, l. 16, insert [bretʃli]
p. 20, foot. 'The early forms of Loughton (Ess), 1062 (15th) KCD 813 *Lukintone*, DB *Lochintona, Lochetona* make it probable that this name has the same history.'
p. 22. l. 13, for 'Ickwell (Mx)' read 'Ickwell (Beds)'
p. 27, l. 15. Wolferton (Nf) and Wolverton (Wo) show the same fluctuations of form.
p. 52, l. 19, for 'stocc' read 'stoc'
p. 83, l. 6, delete the reference to Souldrop (Beds).
p. 106, l. 7, for 'Littlecote below' read 'Littleworth *supra*'
p. 125, l. 4 from bottom, read 'Illington'
p. 130, l. 12. The name *Wyrmhere* is on record in OE in *Widsith* 130.
p. 171, note 1. Mr F. T. S. Houghton makes the very probable suggestion that *Hrisbyri* is to be identified with Risbury (He).

p. 189, 'RASSLER WOOD. Cf. Ratsloe in Huxham (D), 1249 FF *Radeslo*,'

p. 189, l. 12. Mr F. T. S. Houghton points out that the ground in Cofton Hackett (Wo) in BCS 455 is where the keuper marls come up against the Lickey Ridge and that the ground here would certainly be a red mire in its primitive condition.

p. 190, l. 16 from bottom. For the suggested form *medemanhame* an exact parallel is found in the Hampstead Charter of 986 (11th), O.S. Facs, iii, no. 36, *to medeman hamstede*.

p. 191, s.n. *Marlins Grove*. All reference to *Marlingford* should be deleted as the earliest forms of this name make it probable that the true form is *Marding-* or *Marðing-*.

p. 207, l. 1, for '*supra*' read '*supra et infra*'

p. 223, l. 3, for 'Misbourne' read 'Chess'

p. 250, ll. 20, 21. *Lauerkestoke* and Adstock should be under stoc rather than stocc.

p. 251, l. 17, insert 'xv, xvi.' after 'Introd.'

ib. l. 28, insert 'xvi.' after 'Introd.'

p. 266, col. 1, l. 6, for '56' read '55'

p. 272, col. 1, delete the reference to *Ellington* (Nf).

p. 272, col. 2, for '*Ickwell (Mx)*' read '*Ickwell (Beds)*'

ib. insert '*Illington (Nf)*, 125'

p. 273, delete the reference to *Marlingford (Nf)* and to *Souldrop (Beds)*

ib. for '*Medmeuy*' read '*Medmeney (Sx)*'

p. 274, col. 1, insert '*Willey (Sr)*, 85'

ADDENDA

TO VOLUME III

p. 1, AKEMAN STREET. Professor Ekwall points out that the term *Fosse streat* is applied in BCS 1257 to the Roman road west of Bath, near Clifton, and that in BCS 922 we have the phrase *per stratam publicam que ab antiquis stret nunc fos nuncupatur* used of the Fosse Way near Brokenborough (W) in its Bath-Cirencester stretch. It is clear, therefore, that even in early times the Roman road must have been called *Fos(se)* both east and west of Bath and that Akeman Street can only be an alternative name for the stretch east of Bath. He also points out that if the OE pers. name which lies behind Akeman is *Ācemann* we should have expected ME *Ōkeman*. That is the case, but it is possible that one might also have had *Ăkeman* with trisyllabic shortening also. Such shortening is perhaps suggested in the *Alkemanne-strete* form given above, where *lk* is clearly an error for *kk*. We have no precise knowledge of any local pronunciation of the road-name, though it may be noted that there are maps which print the name as *Ackman*. The modern pronunciation with a long vowel may therefore be purely an antiquarian convention.

He also adds a further reference to this road from the Beaulieu Cartulary (90 *b*, 91 *a*) where the name *Ak'manestret, Akermannestret* is used of this road near Shilton (O) and, of much greater interest, calls attention to the use in the Westminster Charter of Edward Confessor, printed in Armitage Robinson's *Gilbert Crispin* (168), of *Akemannestrete* as the name of an unidentified road in London, south of Watling Street (i.e. Oxford Street) and leading to Charing (i.e. Charing Cross). This is a striking example of an early unhistorical extension of the name unless we take this to be the beginning of the great western road which ultimately leads once more to Bath.

From the Stevenson MSS we may note a reference to the Akeman Street, near Piddington (O) and Ludgershall (Bk) found in the *Cartulary of St Frideswide* (ii. 97), *Akemannestrete* c. 1294.

p. 4, ICKNIELD WAY. From the Stevenson MSS we may note the following additional references. In a Bedfordshire Fine of 1209 (cf. BHRS vi. 39) the road is called *Ikenild via*. In the Close Rolls (1274) we have mention of an acre of land 'at *Hykenhilte*' referring to the road near Steeple Morden (C) and of interest as a further example of the use of the name of the road with no addition of *via, strete* or the like. In RH (ii. 445 *a*) we have also a reference to it near Fulbourn and Babraham (C). Stevenson also notes the use of the name *Ekenelde-wey* of an entirely different road in 1371 (Cl) where it applies to the old road from Winchester to Marlborough, near Barton Stacey (Ha).

Mr Schram calls our attention to the use of the term *regia strata vocata Ykenildesthrete* of a road in Dersingham (Nf), which may throw light on the obscure question of the course of the Icknield Way after Thetford.

p. 5, WATLING STREET. To the examples of the extended use of this name we may add *Watlyngstrete* (Pat 1433) used of the road from Ferrybridge to Worksop (Stevenson MSS).

p. 21, YELDEN. Professor Ekwall suggests alternatively that Yelden may be from OE *Gifla-denu*, the valley of the *Gifle*, the name of the settlers of the whole district. (Cf. Introduction xviii and 94 *infra*.)

p. 22, CLAPHAM. Professor Ekwall points out that Skeat is hardly justified in rendering the Middle Danish *klop*, which he apparently took from Kalkar's dictionary, as 'stub, stump.' Kalkar, quoting from an earlier MS dictionary, renders it *klods*, i.e. 'block, lump.'

p. 44. Professor Ekwall calls attention to the survival of *Stachesby* as Stakesby Hall near Whitby and considers that here as in the neighbouring Staxton we have ODan *Staki*. The survival as *Stakesby* is definitely against using that name as evidence for ON *Stakkr* in England, provided that the modern spelling represents the true pronunciation.

p. 70, ELSTOW. Professor Zachrisson in the *Anglo-Norman Influence etc.* p. 147, suggested that this was really *ellen-stow*, 'alder-tree site,' and explained the *aun-* forms as AN spellings. The possibility of such a compound, of which, up to the present, no example had been noted, is shown by the name *apuldorstow* in the boundaries of Ottery in an unpublished Saxon charter communicated to the editors by Mrs Rose-Troup.

p. 103, DUNTON. The early forms perhaps point to an alternative form *Duninga-tun*, 'farm of the hill-dwellers,' *v.* ingatun.

p. 176, l. 14. The actual reading in the text of DB, or rather in an interlineation in the text, is *castellum in Pechefers*. This has often been taken for Peak Forest, but *fers* is an impossible form for such an interpretation. Peak's Arse is an old name for Peak Cavern and *Pechefers* may well be an error for *Pechesers*. The castle is *supra* rather than *in* Pechesers, but one need not perhaps press the meaning of an interlineation too closely.

p. 186, FLETTON. The *fleet* is apparently referred to in a perambulation of 1578 (Hunts NQ iii. 387) where we have mention of a piece of land called the 'Fleete.' Similarly, in the Enclosure Award we have a piece of land called 'the fleets' (*ex inf.* Mr H. E. Norris).

p. 188, CHALDERBEACH. Mr H. E. Norris suggests that the 'diver-birds' were Great Crested Grebes, which used to breed there.

p. 194, BOTOLPH BRIDGE. Mr Bruce Dickins calls attention to the fact that the original dedication of the church here was to St Botolph (Arnold-Forster, *Studies in Church Dedications*, iii. 58) and that he was the patron-saint of wayfarers.

p. 204, l. 4. Mr H. E. Norris points out that the identification of the Abbot's Chair and the Hurstingstone was already made by Brayley in 1808. He adds the further interesting point that the stone can be identified as one coming from quarries at Barnack (Nth).

p. 216, GREAT WHYTE. Mr H. E. Norris reminds us that the streets called Great and Little Whyte are built over the beds of two streams which, until 1854, were open, with houses on their banks. Their line does not seem to have been an altogether straight one. Before he was aware of the original topography of this place Professor Ekwall suggested the possibility of an OE **wiht*, a derivative of OE *wican*, 'to move, to yield.' Such a term might perhaps be used of a bend in a stream.

p. 217, WORLICK. The most probable suggestion is that of Mr Bruce Dickins, that the first element is the pers. name *Wulfric*, cf. Worlaby (L), DB *Vluricebi*.

p. 221, BROADWAY. Mr H. E. Norris points out that the identification is by no means certain for this street was first made in 1888 on the site of the old Bullock Market.

p. 222, NEW SLEPE HALL. On the topographical side Mr H. E. Norris points out that this house took the place of Old Slepe Hall at the *east* end of the town, pulled down in 1848.

p. 226, WARBOYS. Professor Ekwall takes the late Lat. *boscus* to be a Latinisation of OFr *bois*, which was borrowed from some Germanic language in medieval times. The *busc(e)* forms here cited he would take to be loan-words from OScand *buskr*, *buski* or possibly from a lost OE cognate *busc*.

p. 255, EYNESBURY. For the rare OE name *Einulf*, Dr Armitage Robinson has called our attention to the p.n. *Aynolfisfrith*, in the Glastonbury district, found in Abbot Monington's *Secretum* 129b. He also notes the name *Enulf* borne by a discipulus who signs a charter of King Edgar (BCS 1270). This may stand for either *Eanulf* or *Einulf*.

p. 263, PAXTON. Gorham (*History and Antiquities of Eynesbury and St Neots*, Appendix, p. xxiii) quotes from the St Neots Cartulary the phrase *land at Paxebroc* which seems to refer to a brook in Paxton.

p. 265, ST NEOTS. The town takes its name from the obscure saint named Neot whose relics were translated to this place in the late 10th century. It is natural to suppose that he is identical with the Cornish saint of that name.

p. 292, brade. Professor Ekwall calls attention to the field-names *Berebrede*, *Dambrede*, *Milnebrede*, *Yapesbrede*, *Langebrocbrede* found in the Ely Register, f. 115, and used of land in Ditton (C). This suggests that the term is specially an East Anglian one.

p. 294, heolstor. Owing to the rarity of the representation of OE *eo* by ME *u* in Beds (*v.* Introduction, xxvi), Professor Ekwall would prefer to take the first element in this name as OE *hulfestre*, 'plover.' This may well be right.

p. 296, weg. As parallels to *Thefwey* we may note *þeofa-ford* (KCD 611) and *Thefstighe*, nr. Newstead (Nt), *Monasticon*, vi. 474. (Stevenson MSS.)

BEDFORDSHIRE & HUNTINGDONSHIRE

BEDFORDSHIRE

Bedanfordscir 1011 E (c. 1200) ASC, 1016 D (11th) ib.
Beadafordscir 1016 E (c. 1200) ASC

HUNTINGDONSHIRE

Huntandunscir 1011 E (c. 1200) ASC, 1016 ib.
Huntadunscir 1011 C (11th) ASC, 1016 D (11th) ib.
Huntedunscir 1011 D (11th) ASC

For the origin of the names of the counties *v.* Introd.

BEDFORDSHIRE AND HUNTINGDONSHIRE
ROAD-NAMES

AKEMAN STREET

Alkemannestrete (sic) Archaeologia, xxxvii. 434[1]
Akemanestrete c. 1260 Rec. of Bucks xi. 343–8
Akemanstret 1291, 1461 ib.
Akemannestrete 1311 BM

This is the name now commonly given to that Roman road which branches off from the Foss at Cirencester and carries the Bath-Cirencester road to Alchester in Oxfordshire and on to Tring. In the Bucks references the name is used, as Dr G. H. Fowler has shown (*loc. cit.*), of the Icknield Way, on which Akeman Street converges south-east of Aylesbury and with which it is identical for a mile west of Tring. In the last reference it is used of the same road in Totternhoe parish. The names of the two roads have clearly been confused.

It is certain that in explanation of this name we must associate it with the old names for Bath itself, *Acemannes ceastre* (973 A), *Acemannes beri* (972 F) in ASC. There is no evidence that the continuation of Akeman Street from Cirencester to Bath, which is now known as the Foss Way, bore this name in ancient times. Before the Conquest the name *Fos* (BCS 229, KCD 620) is only known to have been applied to what is now called the Foss Way north-east of Cirencester, and it may well be that the first stretch,

---

[1] Stevenson MSS. Bounds of Wychwood Forest (O).

from Bath to Cirencester, was in Old English times called *Ācemannes-stræt* from the fact that it came from *Ācemannesceaster*, the old name for Bath. With reference to the ultimate history of these two names allusion must be made to the excerpts from the early 12th cent. *Annales de ecclesiis et regnis Anglorum* (Cott. Vitell. C viii), printed by Liebermann in *Ungedrückte anglonormannische Geschichtsquellen* (19), which asserts that *Akemannesceastre* was so called from a certain *Akemannus*. The late W. H. Stevenson (*Academy*, April 30, 1887) was inclined to accept the truth of this explanation. It should be added that there is good evidence for OE names in *Āc-*. *Ācwulf* must lie behind Occleston (Ch), DB *Aculvestune*, and the name *Aceman* is on actual record in the mid 12th cent. (Stenton, *Danelaw Charters*, 249). The name *Āca* which lies behind Aketon (Sf), *Acantun* BCS 1289, is further evidence for such.

On this theory it may be assumed that the great road leading out of Bath came to be called after one *Aceman*, the Saxon into whose possession the ruins of Bath passed, and that the name of the road, in the fashion to be noted later under other roadnames, came to be used of the road from Bath throughout its whole length.

The Roman road from Cambridge to Littleport is also sometimes known as Akeman Street (Fox, *Archaeology of the Cambridge Region*, 165). This is probably an unhistorical extension of the name, similar to that noted under Ermine Street, Icknield Way and Watling Street.

ERMINE STREET

> *Earninga stræt* 955 (c. 1200) BCS 909, 957 BCS 1003, 1002 (12th) Proc Soc Antiq (Second Ser.) ii. 49
> *Ermingestrete* c. 1090 (c. 1230) Laws of William 1, c. 1400 Higden
> *Erningestrete* c. 1230 HH, c. 1300 RG
> *Arningstrat* 1251 Ch[1]
> *Arnygestrate* Hy 3 HMC Rutland iv[1]

The clue to the interpretation of this name was discovered independently by the Rev. A. Goodall and each of the editors.

[1] Stevenson MSS.

Mr Goodall points out that it must be associated with Arrington and Armingford Hundred in Cambridgeshire. Arrington is on Ermine Street, just north of the Cam, and its early forms are *Ærningetune, Erningetone* (PN C 15). Armingford Hundred is a Hundred chiefly on the south side of the Cam, which includes the parishes of East Hatley, Tadlow, Croydon, Clopton, Wendy, Shingay, the Mordens, Abington, Litlington, Whaddon, Melbourne, Meldreth, Kneesworth, and Bassingbourn. These parishes form a half-circle round the ford which must once have carried Ermine Street across the Cam, and there can be no doubt that the old name of the ford was *Earningaford* or, as given in BCS 1265, 1266 and 1267 respectively, *Ærningaford, Earmingaford, Earnigaford*. The natural inference from this is that what is now the name of the whole road from London to Lincoln was originally given to that stretch of it which ran through land settled by the *Earningas*, i.e. by *Earn* and his people, and that later, as in the case of Watling Street *infra*, the term was given a wider extension and applied to the whole of the Roman road. In the references to Anglo-Saxon charters given above, the name is used of Ermine Street at Conington (Hu), just before it crosses the Nen to the west of Peterborough, and by Fen Stanton. In the Laws of William the Conqueror it is named as one of the four roads in England which were subject to a special peace safeguarded by heavy penalties, and it is clear that the wider application had already developed. It may be added that in the Charter Roll reference it is used of the road as it passes through Royston (Herts) and in the Rutland MSS as it passes through Papworth (C).

From the peculiar legal privileges of special peace protecting travellers on these roads and their general importance arose the habit, further illustrated under *Watling Street* and the *Icknield Way*, of naming other important roads after them. This habit is commonly ascribed to late antiquarian usage, but there is ample evidence that in some cases it goes back to much earlier times. There is an Irmine Street now used as the name of the Roman road from Gloucester to Silchester. Already as early as the *Polychronicon* we have Higden using the term *Ermingestrete* in his enumeration of the four great roads of Britain, not of Ermine Street as we now commonly know it, but of a road from

St Davids to Southampton, clearly, in part at least, the Glouces-
ter-Silchester road just mentioned. There is also an Irmine
Street (part of Stane Street) on Leatherhead Downs (Codring-
ton, *Roman Roads*, 55). *v.* Addenda.

ICKNIELD WAY

> *Ic(c)enhilde weg* 903 (12th) BCS 601, 944 (12th) ib. 801
> *Icenhylte* 903 (12th) BCS 603
> *Icenilde weg* 956 (c. 1250) BCS 1183
> *Ycenilde weg* 973 (c. 1250) BCS 1292
> *Icenhilde weg* c. 1050 (c. 1250) BCS 479
> *Hykenild* c. 1090 (c. 1230) Laws of William 1
> *Hikenilldes weye* 1279 RH

Whatever be the ultimate etymology of this road-name it
should be made clear at the outset that this is not the name of
a Roman road but of an ancient British track-way and that,
appropriately enough, while the other three great roads, Watling
Street, Ermine Street and the Foss Way, contain in their names
an element loaned from Latin, there can be little doubt that
Icenhylte is purely British in origin, and from the reference in
the charter of 903 it seems that it could be used without
the addition of even such a word as *weg* to indicate the road.

In the pre-Conquest charter references it is used of that
ancient road at Wanborough (W), Hardwell in Uffington, Harwell
and Blewbury (Berks), and Risborough (Bucks). We have no
very early evidence for the application of this name to any of
the extensions of the track-way east or west. For the westward
extension towards Marlborough we might have expected
charter-evidence if the name had been in actual use. For the
eastward extension in the Royston-Baldock direction we have
no charter material at all, but we have the 13th cent. reference
to the road near Chippenham in East Cambs.

From its being one of the four privileged roads of the Laws
an early extension of the use of the name of this ancient road
took place. The series of references given below[1] show that

---

[1] In Alvechurch we have Wm de *Ikenildstrete* in 1319 (Pat), Henry
Ikelyngstrete in 1340 (NI), terra in *Ekelingstrete* in 1535 (VE). In Littleton
we have Nicholaus *Iggengyld* in 1275 (SR) and Ricardus *Ikenild* in 1327 (ib.).
(References due to Mr F. T. S. Houghton.)

as far back as the 13th cent. there was an Icknield Street in
Worcestershire, certainly at Alvechurch and probably also at
Littleton, where it must, in that case, have come to be used for
that road commonly known as Buckle Street, earlier *Buggildstret*
(BCS 126) and *Bucgan stræt* (BCS 1201). This must be the line
of road now called Icknield Street on the maps, and that in its
turn be identified, at least in part, with the mysterious *Rikenild
stret* of which we first hear in Higden's *Polychronicon* ii, 46
(c. 1400) where he speaks of that road as running from St Davids
by Worcester, Birmingham and Lichfield to York. There has
been much idle speculation as to the origin of the form *Rykenild*.
It is clear from the forms given above that it is simply a case of
ME *at there Ikenilde strete* developing to *at the Rikenilde strete*.

The name *Ikenildestreta* is also used for an entirely different
road in the Pipe Roll for 31 Hy 2, where it refers to the old
Roman road from Salisbury to Badbury and Dorchester.

In the present state of our knowledge it does not seem profit-
able to enter into any discussion of the ultimate meaning of the
name Icknield itself. It should perhaps be noted that although
the name occurs in a series of Old English charters in no case is
the MS authority for the form of the name earlier than the
13th cent. Recent work, especially that of Dr Cyril Fox, has
given fresh illustration of the high antiquity and great importance
of this road. *v.* Addenda.

WATLING STREET

Wætlinga stræt c. 880–890 Peace of Alfred and Guthrum, 956
    (c. 1200) BCS 986, 1013 E (c. 1050) ASC
Wæclinga stræt 926 (c. 1200) BCS 659, 944 BCS 792, 975
    (12th) ib. 1315, 986 (c. 1000) O.S. Facs. iii, no. xxxvi,
    1013 C (11th) ASC
Wæxlingga strat 926 (c. 1200) BCS 659
Watlingan stræt 957 (post-Conquest imitative copy), O.S.
    Facs. ii, Westminster ii, 1013 D (11th) ASC
Uueclinca strata 1013 F (12th) ASC
Watlingestrete, Wetlingestrete c. 1090 (c. 1230) Laws of
    William 1

In the above references for this name we have the name
Watling Street applied to the road now known by that name at

Hampstead and Edgware (Mx), Chalgrave (Beds), at the point where it crosses the Ouse, at Stowe Nine Churches and at Weedon (Nth), and (in the ASC) an entirely vague reference. In the last reference given it is used as the name of one of the four great roads of England. The only Old English charter evidence for the country north-west of Stowe Nine Churches through which Watling Street makes its way is the Aston by Wellington Charter (BCS 1315) which shows that by 973 the name was applied to other stretches of the great Roman road from London to Wroxeter. The presumption is that by the time of William the Conqueror it had its present full extension, and later writers like Henry of Huntingdon and Robert of Gloucester explicitly so define it.

As to the origin of the name there can be no doubt that here as in Ermine Street *supra* we have a name of purely local and limited application gradually extended in its use. One must associate *Wætlinga* or *Wæclinga stræt* with the Old English name for the Roman settlement at Verulamium. In Bede this is called *Uaeclingacæstir* (Moore MS) with variant forms *Ueclingua-* (Namur MS), *Uaeclinga-* (Cott Tib C ii), with *Wæclinga-* in the OE Bede (10th). In the *Old English Martyrology* (c. 950) we have *Wætlingaceaster* with an alternative reading *Wealynga-* (c. 900). The form *Wætlingaceaster* is also found in KCD 672, a 13th cent. copy of a 10th cent. charter. It is clear that the same settlers who established themselves on the site of Verulamium gave their name, not only to this site but to the Roman road on which it stood. Soon the name of the road was extended in its application in the same way that we have seen the use of the name Ermine Street developed.

The exact name of these settlers raises difficulties. The evidence set forth above makes it clear that the true form of their name was *Wæclingas* rather than *Wætlingas*, and the possibility of such a personal name and of a later development to *Wætl-* is made certain by two other place-names:—Watlington (O), *Wæclinctun* in BCS 547 and Watchfield (Berks), *Wæclesfeld* in BCS 675. If the patronymic comes from a personal name *Wæcel*, as seems fairly clear, the *c* should be a front *c* and Watchfield shows the phonological development which we should normally expect. It would seem however that in Watlington and Watling Street and *Watlingchester* the front *c* became a back *c*, pronounced as *k*. Such a pronunciation may in part be due to early syncope of the unstressed *e*, bringing the *c* and *l* together, but it was probably influenced also by ready association with the adjective *wacol*, 'wakeful,' with back *c. wæcel* must indeed be interpreted as a derivative from the same stem as *wacol*, with a suffix *-il* which caused mutation of the stem vowel. For such a pers. name cf. OGer *Wachila* (Förstemann PN *s.n.*). Later development of *Wakl-* to *Watl-* is an example of a common confusion of *k* and *t* as in Devonshire Battleford, Gortleigh, Rattlebrook, all with *t* for *k*.

We know nothing of these *Wæclingas*. There is no reason to suppose that they were any more important than any other of the numerous settlers bearing group-names in *-ingas*. It was only because their name had come to be associated with so famous a road that Florence of Worcester started the legend that they were sons of a certain king *Weatla* who made a road right across England. Suspicious as this story is in itself·it is clear that it must go completely by the board when we realise that the name behind the road is not a name with *tl* in it at all.

Its use for the Roman road running southwards from Wroxeter through Church Stretton and along the Welsh border was perhaps inevitable. The name Watling Street was, like that of Ermine Street and Icknield Way, later extended. It is also used of at least one other entirely independent road, viz. the Roman road from York to Corbridge and the Cheviots. *v.* Addenda.

BEDFORDSHIRE AND HUNTINGDONSHIRE RIVER-NAMES

The forms of the river-names, so far as they are found in early documents, are as follows. Their interpretation is reserved for the present, as they can only satisfactorily be dealt with as a whole.

HAIL, R. (lost)

Haile c. 1180 (c. 1230) *Warden* 21 *b*
Hayle 1256 *St Neot* 137 *b*, 1276 RH
Heile, the water called 13th BHRS i. 116

The first of these references is to a stream close to Perry Wood in Great Staughton in Hunts and is clearly the river Kym as it passes through that parish. The second is to the Kym where it joins the Ouse. The third reference is to a stream in Swineshead, clearly the small one to the south-east of the parish which forms a feeder of the Kym. Further, on the banks of the Kym, is Hail Weston, already known as *Heilweston* in 1199 (*v. infra* 275) and Hail Bridge, *infra* 265. It is clear therefore that in *Hail* we have the original name of the river Kym, whose present name is clearly a back-formation from Kimbolton, the chief town on its banks. For such a river-name we may compare the Hayle R. (Co), *Hegel* in 1125 in WMP.

IVEL, R.

> *Gifla* (gen. pl.) c. 1150 BCS 297
> *Givle* c. 1180 (c. 1230) *Warden* 11
> *Giuele* 1232 FF
> *Yivele* 1294 Ipm
> *Yevelle* 1341 AD vi

This river-name is the same as the Yeo R. (So), mentioned in BCS 695, 894 as *Gifle* and in BCS 546 as *Yevel*. This river-name preserves its full form in Yeovil (So), which stands on its banks, and is present as *Il-* in Ilchester. The name of the Ivel has undergone a similar reduction in Northill and Southill *infra* 93, 96. The first form quoted above is not really the name of the river itself but the name given in the Tribal Hidage to the people who dwelled by it. *v*. Introduction xviii.

LEA, R.

> *Lygean* 895 A (c. 900) ASC 913 A (c. 925) ib., *Lygan* 896 A
> (ib.), *Ligean* 913 D (c. 1050) ib.
> *Leye* 1274, 1294 *Ass*
> *Luye* 1294 *Ass*, 1317 *Ass* (as quoted in BHRS iii. 221)

The river-name is found in the early forms of Luton and Limbury *infra* 155-6, both of which stand on its banks, and also in Leyton (Ess), *Lygetun* in KCD 824, near to where the Lea flows into the Thames.

LOVAT, R. or OUZEL, R.

Louente, Louette, Luuente 1276 *Ass*

(For other forms of this river-name *v.* PN Bk 1.) It is the same as the Lavant River in Sussex for which we have the early reference in ClR (1228) 'aqua quae vocatur la Lovente.' The modern alternative name is a playful derivative of *Ouse*, a would-be 'Little Ouse.' In BHRS (v. 175) an alternative form of this last name is given, viz. *Whizzle* Brook.

NEN, R.

Nyn 948 (c. 1200) BCS 871, c. 960 (c. 1200) BCS 1129
Nén c. 1000 Saints
Nen a. 1022 (c. 1200) KCD 733, 963 E (12th) ASC
Nien c. 1000 Hist. El.
water of Nene 1549 FF
Nean c. 1660 Moore, *Map of the Great Levell*

This river-name is identical with that of the stream now called Rhee, which gave its name to Neen Savage and Neen Sollars in Herefordshire. This is *Nen* in BCS 1007 and KCD 952.

OUSE, R.

Usan (acc.) 1010 E (c. 1200) ASC
Use 1276 *Ass*
Hus 1276 *Ass*
O(u)se 1287 *Ass*
Ousee 1334 Ipm

SAFERON, R. (lost)

aquam que vocatur Seuerne 13th *Harrold* 31 *b*
Sauerna ib. 32

Dr Fowler first called attention to this 'Severn' stream which, from the passages in the Harrold Cartulary, was clearly a stream in Bedford Borough. He has supplied a passage from the Bushmead Cartulary (50 *b*) which fixes its course more precisely. There we have reference to a *villa* in All Hallows Parish which abuts on the water called *Seuerne*. He shows that this must be the stream shown on Speed's map rising in the north-west near the Friars, flowing past All Hallows and making its way down to the Ouse in a south-easterly direction. For a further note upon this *Severne Diche*, *v*. VCH iii. 8, n. 10. The Rev. F. W. Breed calls attention to the name *Saferondiche* applied to this stream (Farrer, *Ouse's Silent Tide*, 220) and the tradition that the ditch was so called from the colour of its waters, due to the iron found in it. This is of course an impossible derivation. It is clear that the stream is the same as the Severn (OE *Sæferne*, *Sefærn*)[1].

---

[1] In the Newnham Cartulary (116 *b*) we have a grant by one Gilbert, son of Ralph of Renhold, in which the name *Sauernebrygge* occurs. It is impossible to identify the site of the grant, but there can be little doubt that here we have reference to a bridge which crossed this little Severn stream.

BEDFORDSHIRE

Bedford

BEDFORD

Bedanford 918 A (c. 925) ASC
Bydanford c. 1000 Saints
Bedeford 1086 DB
Bedford 1198 FF

'Beda's ford,' the pers. name *Bēda* being the same as that of the great historian and also of the *Bīeda* of the entry in the ASC *s.a.* 501. *Bīeda* is the West Saxon form, *Bēda* the Anglian one. There is a reference in the ASC *s.a.* 571 to a battle fought by Cuthwulf of Wessex against the British at a place called *Bedcanford* (*v.* Introduction xiv). It has often been assumed that this *Bedcanford* is Bedford, but it is clear, apart from any historical difficulties, that *Bedcanford* and *Bedanford* can only refer to the same place if we assume that in OE you might have alternative names of a place, one in which you used the man's ordinary name (here *Beda*) and another in which you used a pet or diminutive form of that name (here *Bedca*). There is no adequate evidence of such a usage. On the phonological side it may be said that an Anglian form *Bedan-* is unlikely in the Parker Chronicle, *s.a.* 918, and to this extent we have further evidence against identifying the two names for we cannot equate WS *Bedanford* and WS *Biedcanford*. The *Bydan-* form in the *Saints* document is however in favour of an earlier WS *Biedan-*, and in that case we must assume that *Bedanford* is a stray Anglian form.

ALDERMANBURY (lost)

Aldemanneþy 1226 ClR
Aldermanbury 1299 Orig, IpmR
Aldermannesbury 1331 QW

This name probably preserves the name of the official residence of the OE *ealdormann* or 'earl' of the county in Saxon times (*v.* burh). Cf. Aldermanbury (Mx).

BATT'S FORD

Batesford 1302, 1340 FA (p)

Sampson and William de *Batesford* are associated with Odell and not with Bedford, so that this suggested identification is very doubtful. On the other hand, no other likely identification of *Batesford* can be suggested.

BRICKHILL

Brichull 1276 *Ass*
Berchull 1276 *Ass*
Bryk-, Brikhull 1287 *Ass*

The history of this name is probably the same as that of Brick-hill (Bk), viz. that it is a compound of the British equivalent of Welsh *brig*, 'top, summit,' and that we have a name in which the British word for *hill* is followed by that word itself. See further PN Bk 31 and note Brickhill (Nt), Dugd. vi. 474 b.

CAULDWELL PRIORY

Caudewelle 1200 FF *et passim*
Kaldewell, Caldewelle 1201 Cur, 1227 Ass *et passim*
Chaldewelle 1224 Pat
Caudwell 1611 Speed

'Cold spring' *v.* **cald, wielle.**

I. STODDEN HUNDRED

Stoden(e) 1086 DB, 1163 P, 1227 Ass
Stodden(e) 1086 DB, 1202 Ass, 1287 *Ass*, 1302, 1316, 1346, 1428 FA
Stodone 1175 P
Stodesden 1177 P
Stotden 1276 *Ass*

v. **stod, denu.** 'Stud-valley' or possibly, if we may assume that the 1276 form preserves unassimilated consonants which have been assimilated in all the earlier forms, we should take the first element to be OE **stott.** The difference of sense would be slight. The site of the meeting-place of the Hundred is unknown.

Bolnhurst

BOLNHURST [bounǝrs] 84 C 9

Bolehestre, Bulehestre 1086 DB
Bollenhirst Wm 1 (c. 1300–25) *Thorney* 7
Bolleherst c. 1150 (c. 1300–25) *Thorney passim*
Boleherst, Bolehurst Hy 2 *AD* (C 5124), 1179 P, 1198, 1200,
    1202 FF, c. 1200 BM, 1202 Ass, 1232 Ch, Cl, 1240 Ass,
    1242 Fees 885, 1247 *Ass*, 1262 FF
Balnherst, Balnhurst 1202 Ass, 1428 FA
Bolhurst, Bolherst 1220 LS, 1247 *Ass*, 1262 FF, 1276 *Ass*
Bonėhurst 1227 Ass
Bolneherst 1240 FF (p), 1271 FF, 1399 Cl
Bolnherst, Bolnhurst, Bolnhirst 1240 Ass, 1276 *Ass*, 1284, 1302,
    1316 FA, 1323, 1330 Ipm, 1346 FA, 1373 Cl, 1390 CS
Bolesherst 1240 Ass
Bollehurst 1242 Fees 868, 1253 FF
Bolehirst, Bolehurst 1247 *Ass*, 1302 FA
Bolenhurst, Bolenherst 1247 *Ass*, 1283 AD i
Bollenhurst 1287 *Ass* (p), 1358 AD i
Bolnehurst al. *Bownest* c. 1550 *Linc*
Bolnest 1493 Ipm
Bonehurste al. *Bolenhurste* 1596 BM

'Bol(l)a's wood.' In the 11th cent. we have two *herst-* forms
and one *i-* form, in the 12th four *herst-* and one *hurst-* form, in
the 13th fifteen *e-* forms, one *i-* form, thirteen *u-* forms, in the
14th two *i-* forms and eight *u-* forms. After that *u-* forms
prevail. *v.* hyrst. It is uncertain whether the personal-name
which forms the first element was *Bola* or *Bolla*, but the 11th
cent. Thorney form makes the latter probable. For *Bola v.*
Redin 85; cf. Bolnoe *infra* 28. The name *Bolla* possibly arose
as a short form of *Bōtlāf*. For the *a-* forms cf. Colworth *infra* 40.

GREENSBURY

Grymesbury 1331 QW

The *Grym* family had a holding in Bolnhurst in 1302 (FA) and
bury is here used in the manorial sense (*v.* burh). Hence
'Grym's manor.' The modern form is corrupt.

MAVOURN

Maverns 1549 LP

Dean

DEAN, UPPER and LOWER 74 J 7

Dene 1086 DB *passim*
Middeldene 1287 *Ass*
Dene juxta Tillebroke 1307 *Ass*
Overdene 1430 Pat
Overdeane, Netherdeane 1539 Dugd vi. 82
Overden, Netherden 1545 LP

v. **denu**. The valley is that of the tributary stream which joins the Til at Lower Dean. *Over* and *Nether* are the usual earlier terms for places now known as *Upper* and *Lower*. Tilbrook is the neighbouring Huntingdonshire parish. Where the settlement of *Middeldene* was it is impossible to say.

HARROWICK

Harewyke, Herewik 1287 *Ass* (p)

The material is poor, but if the form *Harewyke* is to be trusted, Professor Ekwall suggests that the early form might be OE *heargawic* (Anglian *hergawic*). This would mean 'farm of (or by) the sacred groves' (*v.* **wic**, **hearg**). That word is commonly used in the plural if we may judge by the scanty OE charter material. Harrowick stands at the highest point in the parish. For such a hill-situation, cf. Harrow-on-the-Hill (Mx) and Harrowden *infra* 91 and (Nth) and Peperharow (Sr).

HAY WOOD

Netherhay, Netherhaghe 1287 *Ass*

v. **(ge)hæg**. 'Fenced wood.' The second form in -*haghe* shows confusion with the common **haga**, 'hedge.'

Keysoe

KEYSOE [keisou] 84 B 8

Chaisot, Caissot 1086 DB
Kaiesho Hy 2 (1317) Ch
Chaishou 1167, 1176 P
Kais(h)o 1195 P, 1219 FF, 1240 Ass, 1276 *Ass*
Caysho(u), Kaysho(u) Hy 3 BM, 1219 FF, 1227, 1240 Ass,
    1247 *Ass et passim* to 1490 Ipm

Kaissow 1219 FF
Keysho 1227 Ass
Cay(e)sho 1247 *Ass*, 1295, 1296 Ipm
Kayso, Cayso 1276, 1287 *Ass*, 1290 Cl, 1295 Ipm
Gayso 1287 *Ass*
Cayshoo 1368 Cl
Cayssho 1390–2 CS
Caysthoo 1421 IpmR
Keisoo 1579 Cai
Caishoe Eliz ChancP
Caishow 1647 NQ i
Casoe 1767 (*citation by Archdeacon's commission*)

'Cæg's spur of land,' the first of the very numerous examples
of hoh in Bedfordshire names and a clear example of its general
topographical sense, for it lies on a spur of land running down
sharply to the junction of two (unnamed) streams. The pers.
name *Cæg* is not found in independent use in OE, but it is
found in the same compound in Cassio(bury) (Herts), *Cægesho* in
BCS 267. For the weak form of this name, *v.* Cainhoe *infra* 147
and IPN 180. For DB -*t*, cf. DB *Asceshot* = Ashow (Wa).

Knotting

KNOTTING 84 B 6
Chenotinga 1086 DB
Cnottinge, Knottinge, Knottynge 1163 P, 1220 LS, 1223, 1236
    FF, 1242 Fees 868, 1247 FF, *Ass*, 1276 *Ass*, 1284, 1302,
    1316, 1346, 1428 FA
Cnotinges 1224 Bract
Gnottinge 1276 RH
Knotyng(g)e 1287 *Ass*
Knottingge 1287 *Ass*
Knottingg, Knottyngg 1325, 1361 Cl, 1492 Ipm
Knotymg 1337 Cl

'Cnotta's people,' as suggested by Skeat and accepted by
Ekwall (PN in -*ing* 67). *v.* ing. Skeat quotes the parallel of
Cnottingahamm from a Berkshire charter (BCS 895), and Knot-
tingley (Y). The name is not on independent record in OE but is
the equivalent of ON *Knútr*, lit. 'knot.' Cf. also *Knot, Knod* in
Lincoln Assize Rolls of 1202.

Melchbourne

MELCHBOURNE 84 A 7

> *Melceburne* 1086 DB
> *Melcheburn(e)* 1163 P, 1227 Ass, 1241 *Ass*, 1235 Cl *et passim*
>   to 1428 FA
> *Melchburn(e)* 1220 Ass, 1262 FF
> *Meleburne* 1242 Fees 868
> *Melchebourn(e)* 1316 FA, 1341, 1343 Cl, 1351 Ipm, 1358 Cl
> *Milchbourne* 1675 Ogilby

This is a difficult name and no certainty is possible. There is an OE *melsc, mylsc* 'mellow' applied to apples and the NED suggests that this has survived in the dialectal *melch, melsh* which denotes 'mellow, soft, tender' and, of weather, 'soft.' It also suggests the possibility that that word has not been uninfluenced by OE *milisc*, 'honeyed.' It is possible that this word might be applied to the water of a stream which went down specially smoothly or was notably sweet to the taste. For the former we may compare the quotation in the EDD, 'Whiskey went down as *melch* as new milk.' The chief difficulty in this explanation is the uniform *c* or *ch* in the early forms. The only other example of this element that has been noted is in *Melcheheg* in Chalgrave in the 13th cent. *v.* burna. For such a stream-name we may compare Honeybourne (Wo).

Pertenhall

PERTENHALL [paˑtnəl] 84 A 8

> *Partenhale* 1086 DB, 1227 Ass, 1287 QW (*Parva*)
> *Pertinhale* 1179 P, 1276 *Ass*
> *Pertehale* 1191 P, 1201 FF (p), 1241 FF
> *Pertenhal(e)* 1202, 1227, 1240 Ass *et passim* to 1504 Ipm
> *Portenhale* 1242 Fees 868
> *Perteshale* 1276 *Ass*
> *Patenhull* 1284 FA
> *Pertenhall* 1361 Fine
> *Pertenall* 1526 LS
> *Partnale* 16th BRHS viii. 155

Portenhall Eliz ChancP
Peartnall 1605 NQ i
Pertenhill, Portenhill 1675 Ogilby[1].

'Pearta's nook of land' *v.* healh. For this pers. name, not found independently in OE, Skeat quotes the parallel of *Peartan-hal* (Wo) from BCS 1282 and *Peartingawyrþ* from BCS 262, which, it may be added, is almost certainly Petworth (Sx). Further examples of the name are to be found in Partney (L), Bede *Peartaneu*, and Partington (Ch), KCD 749 *Partingtun*.

CHADWELL END

Chawdwell brak 1607 *Terr*

Almost certainly 'cold spring,' from OE ceald-wielle. Cf. the same name in Bk, Ess, Lei, W. For *brak v.* bræc.

ELVEDON (lost)

Elvendon(e) 1086 DB, 1220, 1325 FF
Eluesdon 1163 P
Eluedun 1163–79 BHRS i. 118
Elueden 1179 P, 1202 FF (p)
Eluedon 1220 FF, 1276 *Ass*

The dun or hill of *Aelf* or *Aelfa*. Both these names are on record in OE.

PERTENHALL HOO (Fm)

Le Ho juxta Kynebalton 1301 BM
Hoo 1318 Ch

The farm takes its name from a well-marked hoh or spur of land, *hoo* being a form often assumed by the dat. case of that word in this county. It borders on Kimbolton (Hu).

SHIRDON (lost)

Segresdone 1086 DB
Siredon 1202 FF
Sirendon 1202 FF

[1] The forms in the *Lincoln Registers* have *Pert-* down to 1480 and then two *Part-* and four *Pert-* forms in the 16th cent.

Siresden 1227 Ass
Shiresden 1227 Ass
Siredon 1227 Ass
Schir(e)donho 1227 Ass
Shiredun 1233 Bract

'The dun or hill of *Scīr* or *Scīra*,' assuming, as seems certain, that the DB form is corrupt. For this personal-name, cf. Sherington (Bk) and Sheringham (Nf).

Riseley

RISELEY   84 B 7

Riselai 1086 DB, 1156 P (p)
Rislie 1198 P (p)
Risle(e) 1202 Ass, 1209 FF, 1227, 1240 Ass, 1247 *Ass*, 1253 Ch, 1283 Ipm, 1287 *Ass*, 1326, 1352 Ipm
Rissle 1202 Ass (p)
Risele 1202, 1203 FF, 1227 Ass, 1247, 1276 *Ass*
Riseleg(a) 1219 FF, 1227 Ass, 1255 KS
Risley 1219 FF
Rissele 1255 KS (p), 1276 *Ass*
Rysle E 1 SR, 1280 Ch, 1316 FA, Eliz ChancP
Rys(e)legh 1276 *Ass*
Rissele al. *Rysle* 1319 Ipm
Rysele(y) 1351 Cl, 1489 Ipm
Riesle 1490, 1504 Ipm
Riseley or *Risley* c. 1750 Bowen
Risley 1785, 1798, 1820 Jury

This is probably OE **hris-leah**, 'brushwood-clearing' (as suggested by Skeat, cf. Riseley (Berks, Db)) with later hesitation between shortening of the vowel before the consonant group and retention of the long vowel of the independent ME word *rise*, 'brushwood.' *Risan-leah* from a personal-name *Risa* is just possible, as suggested by Ekwall for Rising (Nf). In that case the original short vowel must have been lengthened in the open syllable and, later, after the loss of inflexional *e*, have been occasionally shortened before the consonant group.

Shelton

SHELTON 74 J 7

Eseltone 1086 DB
Sheltune 1197 FF
Selton 1202 Ass, 1236 Cl, 1284 FA, 1287 *Ass*
Scelton 1202 Ass, 1302 FA
S(c)helton 1242 Fees 893, 1243 FF *et passim*
Sylton 1276 *Ass*
Schylton, Schilton 1276 *Ass*, 1282 Ipm
Selton al. *Shelton* 1292 Ipm

Skeat's suggestion is that this should be interpreted as from OE *scelf-tūn*, 'farm on shelving or sloping ground,' with the same first element that is found in the unidentified *Scelfdun* of BCS 264. This is the same as the element scylf discussed in EPN. Such an interpretation fits the situation of this place and of Shelton in Marston Moretaine *infra* 80. The only difficulty lies in the early and complete loss of the *f*, but it is clear that this is not a serious one, for Shelton (Sa) has *Sultone*, *Schelton* and *Chelton* as its early forms in the 11th and 12th cent., forms with *f* first appearing in the 13th (cf. Bowcock, PN Sa). So also there can be little doubt that Sheldon near Chippenham (W), though the early forms show no *f*, must contain this element as also Sheldon (Wa). Both are names of well-marked hills. For Shelton (St), Shilton (O, Wa) we have forms with *f* and similar topography and the same is true of Shelley (Sf, Sx, Y). Shelton (Nf, Nt), Earl Shilton (Lei) show forms similar to the Beds names. Derivation from *scelf* might conceivably suit the Norfolk place, and it will suit the Leicestershire name, but it is distinctly not applicable to the Nottinghamshire one. On the whole it would seem that *scelf, scylf* denote sharply falling ground rather than that which shelves gently. Ekwall points out (PN in -*ing* 40) that such a sense would suit Shelve (K) and similarly it applies to Shell or Shelve (Wo). Shelfanger (Nf) is an interesting example of a compound in which the second element (*v.* hangra) expresses the same idea as that suggested for the first.

Little Staughton

LITTLE STAUGHTON [stɔ·tən] 84 B 9

Estone (sic) 1086 DB
Stottun 1163 P
Stoctuna 1167 P, 1205 FF
Stocton 1195 P, 1227, 1240 Ass
Sto(c)kton 1241 FF (*Parva*), 1247 *Ass*
Stottone 1302 FA (*Parva*)
Stoutone 1346 FA (*Parva*)
Stowghton, Stoughton 1390–2 CS, 1394 Cl, 1490 Ipm

'Enclosure made of stocks' *v.* stocc. 'Little' in contrast to Great Staughton just over the border in Hunts.

HILL FARM (6″)

Churchehulle 1247 *FF*

The farm is just to the west of the church, which stands on a hill.

WICKEY

Wicheweia 1208 BM
Whikey 1518 VCH iii. 166

The interpretation is uncertain owing to the paucity of forms. The name would seem to be a compound of **wic** (the reference perhaps being to the dairy-farm of the manor[1]) and **weg**, the whole name meaning 'road to the dairy-farm.'

Swineshead[2]

SWINESHEAD [swinzhed] 84 A 8

Suineshefet 1086 DB
Swyneshaued 1209 *For* (p)
Swynesheued, Swines- 1209 *For* (p), 1272 FF, 1285 FA, 1293 Cl, 1294 FF, 1328 Ch, 1428 FA
Swyneheued, Swineheued 1247 *Ass*
Swinehefd 1247 *Ass*
Swynyshed 1525 BHRS ii. 52

---

[1] Cf. *Aldwik* in Little Staughton (*FF* 1206).
[2] Formerly in Hunts.

Swaneshed 1526 LS, 1535 VE
Swanneshed 1542 BM
Swanshed 1549 Pat
Swanshedd al. *Swyneshedd* 1585 FF
Swanshead al. *Swinshead* 1595 FF
Swaynshead 1630 BM
Swinhead 1675 Ogilby
Swinshead 1765 J

There is little doubt that we should take this p.n. to belong to the group of names which includes Swineshead (L, Wo), Shepshed (Lei), Gateshead (Du), and Farcet (Hu), derived from the names of animals. Cf. Manshead *infra* 112. It is improbable that these names were given from fancied resemblance to the shapes of the heads of those animals and one must rather believe with Bradley that 'these names point to a custom of setting up the head of an animal, or a representation of it, on a pole, to mark the place for public open-air meetings' (*Bradley Memorial Volume*, 101). Probably owing to the unpleasant associations of the name, an attempt was made in later times to alter the name and to associate it with *swan* or *swain*.

SWINESHEAD WOOD

    Swinesheved (*boscus*) 1232 Cl
    Swyneheved (*foresta*) 1279 RH

TARBAGS

No early form of this name is known. It is just possible that it contains the personal name *Terebagge* (= tear-bag), found in the 1227 Assize Roll, but unfortunately with no indication of the part of the county to which its holder belonged.

Yelden

YELDEN [jeldən] 84 A 6

    Giveldene 1086 DB, 1228 FF, 1247 *Ass*, 1273 Ipm
    Giveldon, *Gyueldon* 1220 LS, 1247 *Ass*
    Gieuleden 1227 Ass
    Chivelden 1242 Fees 884
    Gyuelden(*e*) 1247 *Ass*, E 1 KS, 1252 FF, 1276, 1287 *Ass*, 1302,
        1346 FA

E

Yueldene 1315 Ipm
Yevelden(e) 1316 FA, 1317 Ipm, 1319 Cl, E 3 Ipm, 1339 AD
 vi, 1461 IpmR, 1660 HMC App. vii
Yelden 1390–2 CS, 1780–1830 Jury *passim*
Yealdon c. 1600 *Linc*
Yeavelden 1660 HMC vii
Yelding 1675 Ogilby
Yeilden 1765 J

This valley (*v*. denu) takes its name from the stream which flows through it. That is now called the Til, but that name, like that of the Kym into which it later turns, is, as we have seen (*supra* 8), only a late and erroneous formation. There can be little doubt that the stream once bore throughout its length the same name as the Ivel (*supra* 8), another tributary of the Ouse. The early forms of the name of that river given above will explain all those of Yelden. For the loss of *v*, cf. Northill and Southill *infra* 93, 96. *v*. Addenda.

YELDEN SPINNEY

le Spinye E 3 Ipm
Self-explanatory.

(*Detached part of Stodden Hundred*)

Clapham

CLAPHAM 84 E 7

Cloppham 1060 (14th) KCD 809, 1253 BM, 1402 AD iv
Clopeham 1086 DB
Clopham 1163, 1167 P, 1220 LS and so *passim* to 1535 VE, except for
Clapham 1247 *Ass*, 1579 AD v
Cloppham 1253 BM, 1582 AD v
Clappam(e) 1542 Ipm, 1545 AD vi
Clappham 1582 AD v
Clopham 17th BHRS viii. 145

One cannot get much further with this name than Skeat's suggestion that the first element is a lost OE word allied to the Middle Danish *klop*, 'stub, stump.' As this element is not dealt with in EPN it may be well to set forth in full the evidence for its presence in English p.n. so far as it has been gathered. The places in question, with the earliest forms noted, are Clapham (Sr), BCS 558 *Cloppaham*, (Sx) DB *Clopeham*, Clapton (Nth) BCS 1061 *Cloptun*, (Mx) 1345 *Clopton*, Clopton (Wa) KCD 666 *Cloptun*, (Berks) 1316 *Clopton*, (C) DB *Cloptun*, (Sf) DB *Clopetuna*, Clapcote (Berks) DB *Clopcote*, (W) 1428 *Clopcote* and Clophill *infra* 146. In addition we must take account of the unidentified *clopæcer* and *clophyrst* (BCS 1282) and *clophangra* (ib. 508). On the topographical side it may be noted that two of these last three compounds favour the association of *clop* with woods. Apart from this, however, examination of the sites of the places in question has not proved helpful, and is hardly likely to, considering the suggested meaning of the term. In the case of the two Claphams and one of the Cloptons we have forms which suggest the possibility of a genitive plural of the word in question, in the others we seem to have a compound of the ordinary type. For Clapham (Sr) Skeat suggests the meaning 'enclosure of stubs or stubby ground,' but if that were the sense we should have expected an OE form in *hamm* and traces of later *hom(m)*. Such forms are not found for either of the Claphams. Alternatively we are driven back on the ordinary *ham* and must interpret the name as 'homestead of the stumps,' i.e. one marked by such, though genitive plural compounds of this type do not seem natural. The other names would furnish less difficulty, they might be interpreted as tun or cot by some prominent stump of a tree or even 'made of logs,' but really we are very much in the dark as to the meaning of these names. Ritter (128) would connect these names with a Germanic **kluppa-* 'rock' and compares Germ. *Klopf*, *Klopp* as farm-names, but the interpretation would still remain difficult. *v.* Addenda.

It will be noted that these names show a tendency to un-rounding of *o* to *a* and that in some of them the process has been fully carried out (cf. EDG § 83 and Wyld, *Colloquial English*, 240). Attempts have been made to associate these names with Clapham (Y), but, from DB *Clapeham* onwards, this name has uniformly *a*, and it must rather be connected with the pers. name *Clapa* recorded in *Osgod Clapa*.

Milton Ernest

MILTON ERNEST [a·nist] 84 D 7

Mildentone 1086 DB
Middeltone 1086 DB and *passim* to 1372 IpmR
Milton 1372 IpmR

The manorial addition is found as *Erneys, Erneis* (1291 NI, 1372 IpmR), *Ernys* (1334 Ipm, 1765 J), *Herneys* (1372 IpmR), *Harnes* (1526 LS, 1535 VE, Eliz ChancP), *Harneys* (1535 VE), *Ernesse* (17th NQ i), *Earnis* (1780, 1791 Jury).

'Middle farm' *v.* middel, tun. Perhaps so called because midway between Clapham and Sharnbrook as one moves up the Ouse. Robert, son of *Ernis*, is associated with Milton as early as 1227 (FF). The normal development would have been to *Arnes*. The modern form is in part a spelling one and in part due to the influence of the pers. name *Ernest*.

MILTON MILL

Molendinum de Middeltone 1279 RH

Oakley

OAKLEY olim [ɔkli] 84 E 6

Accleya 1060 (14th) KCD 809
Achelai 1086 DB
Achelea 1174 P
Akelai, Akelay 1176, 1179 P
Aklye 1220 LS
Akle 1227 Ass, 1247 *Ass*, 1279 RH, 1346, 1428 FA
Acle 1236 FF, Cl, 1242 Fees 868, 1247 *Ass*, 1267 Ch, 1276
    Ass, 1291 Tax, 1292 Cl, 1301 Ipm, 1302 FA, 1307 *Ass*
Ocle 1247, 1276, 1287 *Ass*, 1291 NI, 1316 FA, 1414 IpmR
Okle(e) 1276 *Ass*, 1328 Cl, 1344 Ipm, 1346 FA, 1378 IpmR
Okele 1390–2 CS, 1428 FA
Ocley 1402 AD iv, 1538 BM
Occley 1535 VE
Ockley 1574 BM
Ockeley 1605 NQ i
Ockley or *Oakley* c. 1750 Bowen

'Oak clearing' (*v.* ac, leah) with later development of *ac* to *oke*, followed by shortening in the consonant-group. In recent times the long vowel has been restored under the influence of the independent word. The history of Oakley (Bk) is similar.

OAKLEY HILL

atte Hulle de Clapham 1276 *Ass*

This identification is probable but not certain. It lies near the border of Clapham parish.

II. WILLEY HUNDRED

Wilge, Wilga 1086 DB
Wilie, Wilye, Wylie, Wyly, Wylye 1166–7 P, 1202, 1227, 1240 Ass, 1276 *Ass*

v. welig. It is clear that, though the actual site is unknown, the Hundred must have met by some well-known willow-tree. Trees often marked the site of such meeting-places (cf. Mawer, *PN and History*, 23).

Half-Hundreds

BUCKLOW

Bochelai, Buchelai 1086 DB
Bukelawe 1189 P, 1202 P, 1247 *Ass*
Bukeslawe 1202 Ass
Buckelawe 1227 Ass, 1287 *Ass*
Bucklowe 1227 Ass
Buckslowe 1227 Ass
Bockelowe 1276 *Ass*
Boclouwe 1284 FA
Buckelowe 1287 *Ass*, 1316 FA

'Bucca's hill or barrow' *v.* hlaw. Cf. also Bucklow Hundred (Ch). This half-hundred included Stagsden, Bromham, Stevington, Biddenham, Ravensden, Pavenham and Bletsoe. The site of the meeting-place is unknown, the only hint that we have as to its whereabouts is the mention in the 1307 Assize Roll of a Simon de *Buckelawe* who probably came from this place. He is mentioned there in connexion with Oakley, which is contiguous to Bromham, Stevington and Pavenham. The half-hundred is now absorbed in Willey Hundred.

Biddenham

BIDDENHAM [bidnəm] 84 F 7

> *Bideham* 1086 DB, 1219, 1228 FF, 1242 Fees 885, 1247 *Ass*,
> 1315 Ipm, 1347 Cl
> *Bidenham* 1086 DB, 1202 FF, 1227 Ass, 1232 Ch, 1276, 1287
> *Ass*, 1302 BM, Ipm, 1316 FA
> *Beydenham* 1240 Ass
> *Budeham* 1247 *Ass*
> *Bydenham* 1276 *Ass*, 1302 FA, 1348 Ipm, 1373 Cl, 1390-2
> CS, 1428 FA
> *Bedenham* 1276 *Ass*, 1461 IpmR, 1535 VE
> *Budenham* 1276 *Ass*
> *Bedenham* al. *Biddenham* 1312 Ipm
> *Bydyngham* 1362 Cl
> *Bedynham* 1428 FA
> *Bydnam* 1512 FF, 1535 VE
> *Byddyngham* 1535 VE
> *Bignam* c. 1550 *Linc*

'The ham of Byda,' this being a personal name found also in
Bidwell *infra* 128, Biddles (Bk) and Bidford (Wa).

BIDDENHAM FORD

> *Reyford* 1279 RH
> *Forde* 1279 RH (p)

The curious first form must be from ME *attereyford* for *at
there ey-ford*, 'at the river-ford.' For such forms *v.* EPN *s.v.*
æt. In *Newn* 155 this ford would seem to be referred to as
Kyngesford.

Bletsoe

BLETSOE 84 C 7

> *Blechesho(u)* 1086 DB, 13th AD ii, 1302 FA
> *Blacheshou* 1086 DB
> *Blechisho* 1195 P
> *Blec(c)hesho* Hy ii *AD* C 5124, 1199, 1206, 1219 FF, 1227
> Ass, 1242 Fees 885, 1247, 1253 FF, 13th AD ii, vi, 1302 FA
> *Blettesho* c. 1230 WellsL, 1280 Cl

Blechenesho 1247 *Ass*
Blethesho 1276 *Ass*, 1303 *KF*
Blekenesho 1276 *Ass*
Blethenesho 1276 *Ass*, 1360 Fine
Bletnesho(o) 1276, 1287 *Ass*, 1290 Ipm, 1316 FA, 1330 Cl,
    1350, 1352 Ipm, 1355 BM, 1360 Ipm, 1390–2 CS, 1483
    IpmR, c. 1550 *Linc*
Blettisho 1276 *Ass*
Bletenysho 1276 *Ass*
Bletesho 1287 *Ass*
Bletthesho 13th AD vi
Bletteneshoo 1368 Cl
Bletsho 1376 Ch
Bletneso 1509–10 LP
Bletesoo 1526 LS

This name was clearly *Blæcceshoh(e)*, 'the hoh of *Blæcc*' in
the first instance. This pers. name is the strong form corresponding to *Blæcca*. It does not occur independently, but is certainly
related to the early medieval name *Blac*. It is also found sporadi-
cally in Bletchley (Bk) and is normal in medieval forms of
Bletchingdon (O). The later development presents two difficulties, (i) the appearance of *t*, which has persisted, and (ii) the
temporary development of forms with a medial *n*. In explanation of the former may be quoted Pitsea (Ess) DB *Piceseia*,
Titsey (Sr) DB *Ticesei*, Whitsbury (W) 1157 *Wicheberia*, in all
of which earlier *c* (pronounced as *ch*) has become *t* before a
following *s* (cf. Mawer in *Modern Language Review*, xiv. 342, and
Zachrisson in *Studier i Modern Språkvetenskap*, viii. 126). In all
these names the difficult and unfamiliar combination [tʃs] has
given place to the simpler [ts]. For the intrusive *n* Mr Bruce
Dickins suggests the existence of an alternative form *Blæccan-
hoh*, with the weak form of the pers. name. Then by a natural
confusion the *s* from the strong form of the pers. name was
introduced into the one from the weak form, and we got the
resulting *Blechenesho* (or *Bletnesho* when this is combined with
the sound-change already noticed). The use of the weak form
probably arose from the same cause as the development of the *t*,
viz. the difficulty of the sound-combination involved in the use

of the genitive of the strong form. The late appearance of forms in *n* seems to preclude what would otherwise be the simplest explanation of the name, namely that it represents an OE *Blæcneshoh*, and contains a diminutive pers. name formed from *Blæc* by the addition of an *n*-suffix.

BOLNOE (lost)[1]

Bolnoh 1219 FF
Bolnho 1276 *Ass* (p)
Bolonho 13th AD ii
Bolho ib.
Bollinho 13th AD iii

'Bol(l)a's hoh,' with the same pers. name (possibly of the same man) that is found in Bolnhurst *supra* 13, three miles to the east.

GALSEY WOOD

Galisho 1276 *Ass* (p)

The only suggestion that can be offered is that from the OE adj. *gāl*, 'wanton,' a nickname was formed and that *Galisho* is 'Gal's hoh' with shortening of the vowel. Cf. Galsworthy (D).

Bromham

BROMHAM [brumǝm] 84 E/F 6

Bruneham, Brimeham 1086 DB
Bruham 1164, 1165, 1173 P, 1220 LS, FF, 1242 Fees 885, 1247 *Ass*, 1258 FF, 1274 Ipm, 1276 *Ass*, 1284 FA, 1287 *Ass*, 1296 Ipm, 1302 FA
Bromham 1227 Ass, 1248 Cl, 1271 FA, 1276 *Ass*, 1286, 1295 Ipm, 1297 Ch, 1316 FA, 1324 Fine, Ipm, 1328 Ipm, 1338 Cl, 1346 FA, 1361 Cl, 1366 Ch, 1428 FA
Braham 1227 Ass

---

[1] It is probable that this is *Bourne End*. In Jefferys' map this is marked as *Bone End*. This is fairly common as a corruption of *Bourne End* (cf. Wootton Bourne End *infra* 87) but the name *Bourne* End is singularly inapplicable to the site as it is away from streams. On the other hand, it is on a *hoh*, *Bone End* may be from *Bolnoe End* and *Bourne End* be a 'respectable' corruption of it. Cf. Bone Hill (Wo), 1262 *Bolenhull*.

Bramham 1228 FF
Brumham 1262 FF, 1287 *Ass*, 1576 Saxton
Brunham 1276, 1287 *Ass*, 1291 Tax
Brumbham 1276 *Ass*
Brynham 1276 *Ass*
Broham 1278 Fine
Bronham 1338 Fine
Broam 1360 Ipm
Brounham 1361 Fine
Burnham 1361 Cl

It is evident that in this name there has, from an early date, been hopeless confusion. Skeat was probably right in suggesting that the earliest form was *Brūnan-ham*, 'ham of Bruna,' the second form in DB showing the common error of transcription *im* for *un*. The curious and persistent loss of *n* in the 12th and 13th cent. must be ascribed to Anglo-Norman influence, which may well have been strong in a place so near to Bedford and its castle. The *n* being weakened or lost in pronunciation it is not perhaps surprising that the unmeaning first element was then often turned into *Brom-* (*v.* brom), a common p.n. element, which at once gave sense to the name. Forms such as *Brum(b)-ham*, *Bronham* are due to conflation of these alternative first elements, while the *Bram-* forms show the influence of ME *brame*, 'bramble,' which is very often confused with brom in p.n. forms.

Bowels Wood

As this wood is on the border of Stagsden parish, in which Joh. de *Boeles* held two virgates of land in 1242–3 (Fees 885), it doubtless takes its name from that family. There was a *Bowels* manor in Bromham (VCH iii. 46), but no holding of the family is known in Bromham itself.

Bridge End

The 'bridge' is doubtless that known as *Bydynghambrugge* (1362 Cl) and *Bydenhambrugge* (1363 IpmR), which connected the two parishes of Bromham and Biddenham.

SALEM THRIFT

Sarlefryth 1350 Ipm

The name of this wood is too corrupt for us to interpret the first part, though it may possibly be the same as Salome *infra* 246, but the second element is clearly fyrhŏe and is another example of the development of that word to *thrift*, which we find in Marston Thrift *infra* 80.

Carlton

CARLTON   84 D 5
   Carlentone   1086 DB
   Carlinton   c. 1175 BHRS ii. 131 (p)
   Carleton   1198 FF *et passim*
   Cherlton   1227 Ass
   Carlton   1240 Ass *et passim*

v. karla-tun. The most southerly example in England of this distinctively Scandinavian name.

Chellington

CHELLINGTON [tʃiliŋtən] 84 D 5
   Chelinton   1242 Fees 884, 1279 RH
   Cheluinton   1247 *Ass*, 1272 FF
   Chelington, -yng-   1273 Ipm, 1287 *Ass*, 1400 BM
   Chelweton   1273 Ipm
   Chelewynton   1273, 1303 Cl
   Chelwedon   1276 *Ass*
   Chelvynton   1279 RH, 1316 FA, 1324 Ipm
   Chelwyngton   1287 *Ass*
   Chelwenton   1290 Ipm, 1302 FA, 1310 Cl
   Chelewintone   1304 Ipm
   Chelvynctone   1315 Ipm
   Chellenton   E 3 Ipm
   Cheldyngton   1347 Cl, 1348 Ipm
   Chelvington   1348 Ipm, 1367 Cl, 1390–2 CS, 1400 BM
   Chillington   1393, 1399 IpmR, 1765 J, 1806 Lysons

This is, as Skeat suggests, from OE *Cēolwynne*-tun, 'farm of Ceolwynn,' that being a woman's name in OE.

BRIDGE END

It was doubtless here that Wm *de Ponte de Chelwyngton* (1287 *Ass*) lived.

Felmersham

FELMERSHAM [fensəm] 84 C/D 6

Flammeresham 1086 DB
Falmeresham 1086 DB, Hy 2 (1255) Ch, 1224 FF, 1227 Ass,
    1235 Cl, 1236 FF, 1240 Ass
Felmeræsham 1163 P
Faumerisham R 1 Cur
Felmersham 1198 Fees 10, 1205 FF, 1227 Ass, 1247 *Ass*, 1283
    Ch, *et passim*
F(r)amer(e)sham 1201 Cur
Felmeresham 1205, 1207 FF, 1227 Ass, 1240 FF, 1241 Cl,
    1242 Fees 866, 1276 *Ass*, 1316 FA
Falmaresham 1240 Ass
Falmersam 1283 Cl
Feelmeresham 1320 Ipm
Felmesham 1503 Ipm, 1549 Pat
Femsham al. *Femsam* 1549 Pat
Fensham 1568 Cai
Fenerson Eliz ChancP

Skeat derives the first part of this name from a pers. name *Feolu-mǣr*, for which he quotes the parallel of *Fiolomeresford* (BCS 1111) from a Warwickshire charter. He also gives the curiously apt alternative form found in *Feala-mæres broc* (BCS 124) from a Worcestershire charter, referring to the brook crossed by the above-mentioned ford, and it would seem that variant forms such as these explain the otherwise difficult variation between *Flam-*, *Falm-* and *Felm-* forms. It is certainly the case that OE *fe(o)la*, *feala*, which is the same word as this prefix, has a double development in ME, normally to *fele* but occasionally to *fale*, forms already found in the *Poema Morale* (Egerton MS) and the *Owle and the Nightingale* (l. 628). *Feolu-* is rare in pers. names (cf. OE *Feolugeld* and Gothic *Filumar*). Hence 'the ham of Feolumær.'

FELMERSHAM HARDWICK

Hertwic 1240 Ass
Herdwic 1240 Ass
Hardewyke 1276 *Ass* (p)
Herdwyk in Felmersham 1377 Cl

v. heordwic.

PINCHMILL ISLANDS

Pinch (*molendinum*) 1276 RH
molendinum de Pynches 13th (15th) *Newn* 165
molendinum de Pinge ib. 171

RADWELL [rædəl]

Radeuuelle 1086 DB
Radewell(e) 1185 P *et passim*
Rodwell, Radwell Eliz ChancP

'Red spring' *v.* read, wielle.

STOKE MILL

Stokes 1219 FF (in Sharnbrook)
Stokemulne 13th (15th) *Newn* 165
Stokemylnes 1380 IpmR

v. stoc.

Harrold

HARROLD 84 D 4

Hareuuelle 1086 DB
Harewold(a) 1163 P, 1220 LS, 1227 Ass, FF, 1234 Cl, 1237,
  1240, 1256, 1268 FF, 1276, 1287 *Ass*, 1322, 1354 Ipm, 1388
  Cl, 1476 AD vi
Harewuda 1167 P, 1220 LS
Harawald 1194 P
Harewde 1227 Ass
Harewode 1227, 1240 Ass, 1244 FF, 1276 *Ass*, 1312 Ch, 1322
  Cl, 1331 QW
Harwode 1232, 1234 FF
Har(e)wald 1236, 1244 FF, 1247 *Ass*, 1253 Ch, 1287 QW, *Ass*
Harwold(e) 1240 FF, 1346, 1428 FA, 1432 IpmR, 1501 Ipm,
  1528 LP
Harewaud 1240 FF, 1253 Ch
Horewald 1247 *Ass*

Horwold 1276 *Ass*
Harrold 1346 Ipm, 1535 AD v
Harwood 1525 LP
Harrold al. *Yarrold* 1560 VCH iii. 66
Harwood 1610, Speed, 1730 VCH iii. 66
Harwood or *Harrold* c. 1750 Bowen

The second element in this compound is weald, the reference being to the high ground rising up here from the Ouse valley. The first is har, used probably not in its original sense of 'grey' but in its transferred sense of 'boundary' for the 'wold' lies on the border of the county. Skeat takes the first element to be OE *hara*, 'hare,' but the forms in *Hor-*, of which he was not aware, show that this cannot be right. OE *hār* normally becomes *hore* in ME, but early shortening of the vowel in the first element of the compound ultimately led to the development of *Har-* rather than *Hor-*.

DUNGEE FARM and WOOD
    Tungge 1284 FA (iv. 14) (p)[1]
    Dungeo 1485 IpmR
    Duncheo 1501 Ipm

Mr Bruce Dickins suggests OE *Dunninga-hō* (*v.* hoh), which would suit the situation, and compares Bengeo (Herts), DB *Belingehou*, which shows the same phonological development.

HARROLD BRIDGE (6″)
    pontem de Harewold 1276 RH

HARROLD PARK
    parcum de Harewold 1276 RH
    Harwold Park 1515 AD vi

NUN WOOD
    Nonne Wode 1535 VE

The nuns of Harrold Priory had a wood of 20 acres here (RH ii. 329).

[1] The reference is not certain but, as the heirs of Robertus de *Tungge* belonged to the neighbouring parish of Bozeat (Nth), it is distinctly probable.

PRIORY FARM

Takes its name from Harrold Priory.

SANTON BARN

> *Swanton* 1240 Ass, 1244 FF, 1253 Ch, 1287 *Ass*, 1311 Fine,
> 1324 Ipm, Cl
> *Swonton* 1251 BM, 1276 *Ass*, 1311 Fine, 1325 Cl
> *Swenton* 1331 (15th) *Newn* 44
> *Santon* 1765 J

OE *swāna-tun*, 'farm of the peasants,' *v.* swan, tun. Cf. Sannacott (D), earlier *Swanecote*.

TEMPLEGROVE SPINNEY (6")

This takes its name from the Knights Templar, who in 1279 had a holding in Harrold (RH ii. 329).

Odell

ODELL [wʌdəl], [oudəl] 84 C/D 5

> *Wadelle* 1086 DB
> *Wadehelle* 1086 DB
> *Wahelle* 1162 P (p)
> *Wahull* 1222, 1232, 1236 FF, c. 1230–40 *Bodl* (Bucks), 1235
> Fees 460, 1240 Ass, 1242 Fees 891, 1244 Cl, 1245 Ch, 1276
> RH, 1284 FA, 1287 *Ass*, 1302, 1316 FA, 1324 Ipm, 1346,
> 1428 FA
> *Wa(h)ille* 1240 Ass
> *Wadehulle* 1276 *Ass*
> *Wodhull* 1276 RH, 1287 *Ass*, 1296, 1308, 1337 Ipm, 1358
> Fine, 1376 Cl, 1390–2 CS, 1422 IpmR, 1428 FA
> *Wahull* al. *Wadhull* 1304 Ipm
> *Wod(e)hill* 1359, 1382 Cl, 1476 IpmR
> *Wodell* 1485 IpmR
> *Odyll, Odill* 1494, 1502 Ipm
> *Wodhyl* 1503 Ipm
> *Odell* 1526 LS
> *Odell* al. *Woodhill* 16th BHRS viii. 155

'Woad hill' *v.* wad, hyll. It will be noted that all the names with wad given in EPN are compounded with an element denoting a hill. The forms with loss of medial *d* (cf. IPN 111) must be explained as due to AN influence. Such would be very strong seeing that Odell was the head of an important barony, commonly known to historians by its Normanised name of *Wahull*.

BROWNAGE WOLD

> *Brownehegge* 1469 IpmR
> *Brown Hedge* 1826 B

Probably 'brown edge' rather than 'brown hedge' as the wold runs along high ground. *v.* **brun, ecg, hecg**. Inorganic *h* is very common in compounds whose second element begins with an *e*.

GREAT CATSEY WOOD

> *Gatesho* 1240 FF (p), 1254 FF, 13th (15th) *Newn* 159 *b*
> *Gotesho* 1254 FF (p), 1279 RH (p), 1284 Ipm
> *Gottsho* 1276 *Ass* (p)
> *Cotesho* 1276 *Ass* (p)
> *Cattessho* 1367 Cl

This is a difficult name. It is just possible that it may be 'goat's hill' (*v.* hoh). The proper gen. of *gāt* in OE is *gāte*, but if the name is of ME origin, as it may well be, gen. *gātes, gōtes* is quite possible (cf. also Gateshead in PN Nb Du). Shortening of the vowels of *gates* and *gotes* in the trisyllabic compound would explain the later forms. For initial *c*, cf. IPN 114.

NORTHEND (Fm) (6")

> *Northende* 1279 RH (p)

Self-explanatory.

TREVOR (Fm) (6")

> *Tretteford* 1287 *Ass* (p)

There is very little to go upon, but if the identification is correct one may suggest that the name is from OE *treowihte-ford*, 'tree-grown ford.' This adjective is unknown in OE, the usual form being *treowen*, but would be a regular formation, and, just as we have the pair of OE adjectives *stænen, stæniht*, 'stony,' so we may have had adjectives *treowen, treowiht*, or,

with the regular mutation, *triewen, triewiht.* This explanation
would suit the topography. Other similar adjectives in OE are
cærsiht, hǣðiht, hrēodiht, ifiht. Cf. also *Sandithul* in Colworth
(13th).

WOLD (Fm)

(*ad*) *Wold* 1279 RH (p)

Self-explanatory.

YELNOW

Jelnho 13th *Ashby*

OE (*æt þæm*) *geolwan hohe,* 'at the yellow-coloured hoh.'
Local topographical enquiry suggests no reason for this name.

Pavenham

PAVENHAM [peˑətnəm], [pævənəm], olim [peikənəm] 84 D 6

Pabeneham 1086 DB
Pabeham 1227 Ass, 1242 Fees 807, 1247 *Ass*
Papeham 1240 Ass
Pabenham 1240 Ass, 1266 FF, 1276, 1287 *Ass,* 1302, 1316
    FA, 1322 Ipm, Cl, 1344 Pat, 1346 Fine, FA, 1348 Ipm,
    1388, 1427 BM, 1526 LS, 1535 VE
Pabbeham 1247 *Ass,* 1276 *Ass* (p)
Pakenham 1276 *Ass,* 16th BHRS viii. 148
Pabham 1283 Cl
Papenham 1302 FA, 1344 Pat, 1383 Cl (p), 1406 IpmR
Pavenham 1492 Ipm, Eliz ChancP
Patenham 1576 Saxton, 1675 Ogilby
Patinham c. 1750 Bowen
Pavingham 1766 J, 1806 Lysons, 1820 Jury

'Papa's ham.' The phonology of this name is not easy, for
it is difficult to say whether one should take the first element
originally to have been *Paban-* or *Papan-.* Writing in ignorance
of the *Pap-* forms Skeat took the first element to be a pers. name
Paba. The existence of such a name as this is doubtful and, if
we start from *Paba,* the *Pap-* forms can only be explained by
assimilation of the second to the initial consonant, a somewhat

rare development. It is simpler to adopt the reverse process, take the pers. name *Papa* as the original one and explain the *Pab-* forms as due to voicing of *p* to *b* before *n*. For a name *Papa* Skeat gives good evidence under Papworth (PN C) where he quotes *Papanholt* from BCS 596. The only other possible explanation of the *Pap-*, *Pab-* forms is to take them as from alternative hypocoristic names *Pappa*, *Pabba*, showing the same alternative forms with voiced and unvoiced consonants which we find in other such names. The development of *b* to *v* is found elsewhere in Bavington (Nb), Baverstock (W), *Averton* for *Abberton* (Ess), and *v.* Aversley and Abbotsley *infra* 197, 252. The forms with *k* suggest a pronunciation in which dissimilation of *Pap-* to *Pak-* has taken place, while the modern pronunciation with a medial *t* shows assimilation to the following *n*. For some parallels to the consonantal changes, cf. the history of Hepmangrove *infra* 207, and cf. that of Collett (Bk), earlier *Collick*, and Colletts Green (Wo), earlier *Colewyk*.

STAFFORD BRIDGE

aqua de Stafford 1227 Ass

The 'water of Stafford' would seem to have been the name of the stretch of the Ouse by the present bridge. The origin of *Stafford* is presumably the same as that of the well-known town, viz. *stæþford* (*v.* stæþ). 'Bank-ford' or 'landing-place ford' seems at first sight a curious compound. Its justification is probably that the southern bank of the Ouse slopes fairly sharply at this point. This early form incidentally disposes of the tradition that the bridge took its name from the Staffords, lords of the Honour of Gloucester, to which Pavenham was attached.

Podington

PODINGTON [pʌdiŋtən] 84 B 4

Potintone 1086 DB
Podinton, Podynton 1086 DB, 1223 FF, 1227 Ch, 1240 Ass,
   1247, 1276 *Ass*, 1315 Cl, 1346 Ipm
Pudinton 1163 P, 1202 FF, 1227 Ass
Pudington 1227 Bract
Potingdon 1240 Ass (p)

F

Podington, Podyngton 1287 *Ass*, 1302, 1316 FA, 1324, 1388 Cl, 1390–2 CS, 1509 AD v
Pudyngton 1346 FA
Puddington, Puddyngton 1535 VE, 1588 D, 1643 HMC iii, 1765 J, 1780–1830 Jury

'Pudda's or (possibly) Poda's farm' *v.* ingtun. Both *Podda* and *Puda* are well-established OE names.

COCKLE SPINNEY

Cochul 13th AD ii

Cf. Cockhill (So), Cockle Park (Nb), Cook Hill (Wo). All alike are probably named from the domestic bird.

FARNDISH

Fernadis 1086 DB
Farnedis 1086 DB, 1276 *Ass*, 1284 FA
Fernedeis 1176 P
Fernedis 1194 P, 1207 FF
Franedis 1202 Ass (p)
Farnedis(c)h 1227 Ass, 1273 Cl, 1276 *Ass*, 1346 Ipm
Farendich 1247 *Ass*
Ferenedich 1249 FF
Frandich 1250 Fees 1230
Farnedich 1276, 1287 *Ass*, 1302 FA, 1346 Imp, FA
Farnadis 1276 *Ass, Farnadich* 1316 FA
Farndisch, Farndisshe 1322 Ipm, 1329 Cl
Farendissh 1346 Ipm, Cl, Fine
Farredish 1346 Ipm (p)
Faryngdyssch 1390–2 CS
Faredishe 1399 IpmR
Faryndish 1418 IpmR
Frendish al. *Frendich* 1509 Ipm

'Enclosed pasture, overgrown with fern or bracken' *v.* fearn, edisc. Skeat takes the last element to be *disc*, 'dish,' but there is no evidence in OE charter material for its topographical use, and 'hollow' does not describe the site of Farndish. The later forms show that dissyllabic development which arises from the strong trilling of the *r*, or alternatively metathesis of the *r*.

GORERONG (Cottages)

Wrongelond c. 1200 *Ashby* 195
Little Gorewrong 17th VCH iii. 82

It is clear that both elements in this name must refer to the division of the old open field into *lands*, some of which might be crooked (*v.* **wrang**) and between which there might at times be 'gores' (*v.* **gara**), but the exact sense of the compound is not clear.

HINWICK [hinik]

Heneuuic(h) 1086 DB
Haneuuic(h) 1086 DB, 1207 *P*
Henewich, Henewic, Henewik 1166, 1169 P, 1175 P (Chanc Roll), 1177, 1183 P, 1223 FF, 1240 Ass
Hennewic(h) 1167 P (Chanc Roll), 1175, 1179 P
Hinewic, Hynewyk, Hynewik, Hinewike 12th cent. Northants Survey, 1227 Ass, 1242 Fees 884, 1247 *Ass*, 1302 FA, 1312 Ch, 1315, 1322 Ipm, 1346 FA, 1388 Cl, 1428 FA
Hinwick 1235 FF
Henewyke 1276, 1287 *Ass*, 1322 AD iii, 1346 Ipm, 1428 FA, 1456 AD i
Hynnewyke 1346 Ipm
Henwyke 1407 (*juxta Podington*), 1432 IpmR
Hynwyk 1501 Ipm

Possibly 'fowls' farm' *v.* **henn, wic.** Cf. Goswick (Nb). The later forms show the raising of *e* to *i* which is so common before *n*, cf. the history of *hent* and *hint* (NED), while the *a* of the DB forms will on this explanation be due to the typically 'eastern' *æ* for *e* before nasals (*v.* Introd. xxiii).

Sharnbrook

SHARNBROOK [ʃaˑmbruk] 84 C 6

S(c)ernebroc, Serneburg 1086 DB
Shernebroc 1163 P
Scarnebroc 1167 P, 1220 LS
Sharnebroc 1167 P

Sarnebroc 1189 ChR, 1199, 1202 FF
S(c)harnebrok(e) 1247 FF *et passim*

and other similar forms except

Scharneburg 1276 *Ass*
S(c)hernebrok 1276 *Ass*, 1302 FA, 1494, 1502 Ipm, 1617
   HMC iii
Sharbroke 1406 IpmR
Sharmbrook 1785 Jury

'Dung-brook,' presumably from its fouling by the cattle.
v. scearn, broc. Cf. Sharnford (Lei).

ANTONIE FARM

This takes its name from the *Antonie* family who held Colworth Manor in the 18th cent. (VCH iii. 92).

ARNOE

Audenho 1276 *Ass* (p)

The identification is not certain. If correct, the modern form
is corrupt for *Alnoe*, and the name should be interpreted as
from OE *Ealdan-ho(h)e* ('Ealda's hoh').

COBB HALL

Takes its name from the *Cobbe* family, who held a manor here
in the 17th cent. (VCH iii. 91).

COLWORTH

Colingwurth 1242 Fees 886
Colingwrth 1247, 1276 *Ass*
Colyngworth 1276 *Ass*
Coleworth 1276 *Ass*
Colworthe 1279 RH
Culworth 1392 IpmR[1]

'Cola's enclosure' (*v.* worþ), with the same use of *-ing* as in
ingtun.

---

[1] It should be noted that in the 13th cent. we have forms *Calworthe*,
Kalworthe and *Caleworth* in the Newnham Cartulary, which apparently
refer to this place but are very difficult to reconcile with the rest, but cf.
Bolnhurst *supra* 13.

FRANCROFT WOOD

Franecrofte 13th (15th) *Newn* 162 *b*
Franescroft ib. 164

'Frani's croft,' the pers. name being of Scandinavian origin and appearing as *Fræna*, the name of a Viking leader in the ASC *s.a.* 871. For the full history of this name *v.* Björkman, NP 42.

HALSEY (Fm and Wood)

Alfiseia, Alphiseie 13th (15th) *Newn* 163, 164
Alseye, Alsiesheye Cott Vesp E xvii. 238[1]
Alsey 1276 RH (p)
Alsowe 1469 IpmR

'Enclosure of Ælfsige' *v.* (ge)hæg. It should be added however that the identification is not quite certain for in Sharn-brook parish there is also a place called

Haselho 1202 Ass
boscus de Haselhou 13th (15th) *Newn* 162

When we note that these have initial *h* and that *hæsel* does occasionally appear as *halse* in p.n., and, further, that there is a Halsey *Wood* as well as a farm we are perhaps right in assuming that Halsey Farm and Wood may originally have been named differently, the farm being *Alsiesheye* and the wood *Haselho*. This perhaps receives slight confirmation from the form *Alsowe* given above, which looks like a blending of two such names. In connexion with the earliest forms it may be noted that OE *Ælfsige* occasionally appears as *Alfesi* in the 13th cent.

PIPPIN WOOD (6″)

This retains the name of *Pipings* manor (VCH iii. 89).

SHARNBROOK MILL

Molendinum de Sernebroc 1199 Abbr

TEMPLE WOOD (6″)

The Knights Templar had a manor here in 1279 (cf. RH ii. 323).

---

[1] Cartulary of St Andrew's, Northampton.

THE TOFT

Toft(e) 1279 RH (p), 1331 QW
Taft 1804 Smith

v. topt. It is noteworthy that this place shares with Toft (C) the honour of being one of the two most southerly examples of this element in English place-names. While it is not as a rule definite evidence for Scandinavian settlement, for it was a common ME word, it is perhaps worthy of note that it is found in the same parish as Francroft *supra* 41. See further, Introd. xix.

TRIKETS BURY (lost)

Trikatesburi 1202 Ass

This name is worthy of record as an extremely early example of the use of *bury* (*v.* burh) in the manorial sense. The Triket family held the manor of Toft from the 13th cent. onwards (VCH iii. 90).

WOODEND PLANTATION (6″)

Wodend 1469 IpmR

Self-explanatory.

Souldrop

SOULDROP [su·ldrəp] 84 B 5/6

Sultrop 1196 P, 1202 Ass (p), 1227 Ass
Sultorp 1198 FF
Suldrop(e) 1202 FF, 1247 *Ass*, 1254 FF, Hy 3 BM, 1276,
 1287 *Ass*, 1316, 1346 FA, 1390–2 CS, 1469 IpmR, 1526 LS
Sulthorp 1247 *Ass*
Soldrop(e) 1247 *Ass*, 1302 FA
S(c)huldrop 1247, 1276 *Ass*
Sholdrope 1276 *Ass*
Sulthrope 1276 *Ass*, 1291 NI, Tax
Schulthorp 1276 *Ass*
Souldrop 1535 VE

Were it not for the entire absence of an inflexional *e* between the *l* and the *t* one would take this name to be from OE *Sulan-þrop*, with the same pers. name that is found in Soulbury (Bk), Sulham (Berks, Sx), but it is difficult to overlook this. OE *sulh*, *sul*, primarily used of the plough, and then of the furrow made by it, do seem to have been used in a topographical sense, presumably to denote a gully or narrow channel, as illustrated in *sulh* (BCS 994, 1290), referring to the same place, and *sulhford* (BCS 166, 1331), referring to two places. Souldrop lies on the edge of a well-marked depression through which runs a tributary of the Ouse, and used by the Midland Railway for making its way from the valley of the Ouse to the point known as Sharnbrook Summit (337 ft.). The end of the valley is called Souldrop Gap on Bryant's map (1826). There can be little doubt therefore that the whole name means 'thorp by the well-marked valley[1].' For the form of the suffix, *v.* þorp[2].

FORTY FOOT LANE

in le Lane 1302 FA (p)

Self-explanatory.

LEE (Fm)

Leye 1276 *Ass* (p)

la Leye 1302, 1346 FA

'Clearing' *v.* leah, or possibly 'fallow' *v.* læge

Stagsden

STAGSDEN 84 F 5/6

Stacheden(e) 1086 DB, 1198 FF, 1202 Ass, 1204, 1219 FF, 1227 Ass, 1235 Ch, 1242 Fees 807, 1247 *Ass*, 1276 RH, 1283 Ipm, 1284 FA, 1287 *Ass*, 1302, 1316 FA, 1366 Cl, Fine, 1390–2 CS, 1428 FA, 1430 IpmR, 1494 Ipm

Staggeden 1183 P (p)

Stachesden(e) 1196 FF, 1227, 1240 Ass, 1247, 1276, 1287 *Ass*, 1310 Fine, 1311, 1334 Ipm, 1346 FA, Cl, Fine, 1358 Ipm, 1385 Ch, 1388 Cl

Stachenden 1220 LS (p), 1225, 1242 FF

Staggesden 1228 Cl

---

[1] This explanation requires the deletion of the reference to Souldrop in PN Bk 83.

[2] A further example of this word *sul* is due to the kindness of Mr F. T. S. Houghton. Southstone in Clifton-on-Teme (Wo) is *Sulstan* (c. 1205), *id.* (1308), *Soulston* (1327), *Solughstone* (1353), *Sulston* (1535). It is the name of an isolated rock in the deep-cut valley of the Teme.

Stakeden 1235 Fees 460, 1276 *Ass* (p)
Stachedon 1247 *Ass*
Sta(c)kesden 1276 *Ass*, 1346 Fine
Stageden 1276 *Ass* (p), 1526 LS, 1610 Speed
Stachden 1535 VE
Stackedene 1553 NQ i
Stagsden or *Stachden* 1766 J

It is easier on the whole to explain this name if we start from the assumption that *Stak-* is the first syllable than if we begin with *Stag-*, for not only are the *Stach-* forms much the earliest and much the most frequent in early times, but it is also easily explicable that *Stak-* should be voiced to *Stag-* before *sd* (aided by association with the common word *stag*), while it would be very difficult to find a reason for the change from *Stag-* to *Stak-*. This presumption is supported by the fact that while we have no evidence for a pers. name *Stagge*[1] we do know that there was an ON pers. name *Stakkr*, really a nickname, denoting a person as hefty as a 'stack' (*v.* Lind, *Personbinamn* s.n.) which is found in Staxton (Y) and in a lost *Stachesbi* in that county (*v.* Lindkvist, *PN Scand. Origin*, lxiii). The existence of other names in this Hundred which certainly contain a Scandinavian element make such a derivation possible. The familiar alternation in English pers. names between strong and weak forms probably led to the use of a genitival form *Stache(n)* in place of the regular *Staches* (cf. *Stachenden* supra). Alternative forms with and without voicing of the *k* seem to have been in use till quite late, cf. Cople and Moggerhanger *infra* 89, 91. *v.* denu. *v.* Addenda.

ASTEY WOOD [aˑsti]

Estey 13th (15th) *Newn* 10
Estheie ib. 17 *b*
Arsty Wood 1826 B

'East enclosure' *v.* (ge)hæg. It is on the eastern boundary of the parish.

BURY END

The site of Burdelys Manor (BHRS viii. 7). *v.* burh.

---

[1] We may note however the mysterious *to stacg inwicum* (BCS 758) which may be for *stacgingwicum* as actually amended in KCD 1131.

DILWICK (lost)

Direwic 1202 Ass
Dilewic, Dilewyk, Dylewyk, Dilewik 1219, 1242, 1255, 1262
    FF, 1325 Cl, 1335 Ipm, 1336, 1388 Cl
Dylewick, Dylewyck 1254 FF, 1287 *Ass*
Delewik 1274 Ipm, 1432 IpmR
Dilwyke 1276 *Ass*
Dillewyc 1276 *Ass*
Dylywyke 1287 *Ass*
Dilik Park 1812 BHRS viii. 5

The first element is probably OE *dile*, 'dill,' but used also
dialectally of certain species of vetch (NED). Hence, 'dill
farm' (*v.* wic). Ekwall finds this element also in Dilworth (La).

DUCKSWORTH

Dukesworth 1276 *Ass* (p)
Dokesworth 1316 FA (p)

This name would seem to be the same as Duxford (C), DB
Dochesuuorde, and to mean 'Duc's enclosure' (*v.* worþ), the
pers. name *Duc* being inferred from such p.n. as Duxbury (La)
and Dukesfield (Nb). The weak form *Duca* was still used in
the 13th cent. in the form *Duce*. As the Bedfordshire forms are
derived entirely from pers. names it is probable that this name
is really manorial and originates from a tenant who came from
the Cambridgeshire village of the same name.

HANGER WOOD

le Hangre 13th (15th) *Newn* 144
v. hangra.

HOW WOOD (6")

la Hou 1276 (p)
v. hoh.

PEARTREE (lost)

Peretre 1347 Cl, Ipm
Peartree 1812 *Award*

Land called so from a prominent tree on it.

WHITE'S WOOD (6")

The Burdelys Manor in Stagsden was held in 1346 by John le White (BHRS viii. 7).

WICK END

> la Wykhend 1279 RH (p)
> atte Wyche 1346 FA (p)

'The part of Stagsden lying by one of its dairy-farms' v. wic.

Stevington

STEVINGTON [stefən] [stivintən] 84 E 6

> Stiuentone, Styuenton 1086 DB, c. 1129 BM, 1237 FF, 1242 Fees 873, 1276 Ass, 1316 FA
> Stiuiton 1196 P
> Stiuinton 1227 Ass, 1242 Fees 867, 1276 Ass, 1282 Cl
> Stiuington 1247 Ass, 1280 Ch, 1287 Ass
> Stiuingdon 1276 Ass
> Steventon 1284, 1346 FA, 1384 Cl, 1390–2 CS
> Stevintone, Stevyntone 1302, 1346 FA
> Stevyngton 1315 Cl, 1428 FA, 1454 AD iii, 1515 LP, FF
> Styvyngton 1349 Cl, 1350 Ipm

'Styfa's farm' v. ingtun. The name Stȳfa is not on record, but it may be inferred from the name Stȳfic postulated for the early forms of Stetchworth (C) by Skeat (PN C 27) and is allied to the Stūf of ASC s.a. 514, both having a long vowel. Allied to these names is the Stybba which lies behind Stibbington (Hu) infra 197. That however, as shown by the gemination of the consonant, must have a short vowel. All these names alike go back to a Germanic stem stŭb- which lies behind the ordinary word stub, 'stump' (NED s.v.).

THE HOLMES

> Holm 1279 RH

v. holmr. The 'holmes' are small islands in the Ouse.

PICTS HILL [pikshil]

> Pikeleshill 1227 Ass
> Pykeshulle 1227 Ass, 1235 Ch (p), 1236 FF (p)

Pixhull 1227 Ass, 1236 FF, 1279 RH (p)
Pikeshul(le) 1242 Fees 867 (p), 1302 FA
Picks Hill 1766 J, 1787 Cary

'Picel's hill' or, if we must not lay too much stress on the first form, 'Pic's hill.' *v.* hyll. This name *Pic, Picel* is found in Pitchcott, Pitstone (Bk) but with palatalised *c*. Here and in Pickenham (Nf), Pickwell (Lei), Pickworth (L), Pickthorne, Pickstock (Sa) we have a non-palatalised *c*.

WOODCROFT

Wodecroft 1350 Ipm

Self-explanatory.

Thurleigh

THURLEIGH [θə·lai'] 84 C 8

(*La*) *Lega* 1086 DB, 1202 FF *et passim*
(*La*) *Leye* 1220 LS *et passim*
(*La*) *Legh* 1227 Ass *et passim*
de la Releye 1287 *Ass* (p)
de la Reyleye 1287 *Ass* (p)
Relye 1309 *Cor Reg*, 1394 IpmR, 1402 AD iv, 1427 BHRS ii. 44
Thyrleye 1372 IpmR
Relegh 1387 Cl
Reliegh 1393 Cl
Rely c. 1400 *Harrold* 51 *b*
Thurle 1461 IpmR
Thyrlee 1483 AD iv
Thurley al. *Relegh* 1518 LP
Thurly(e) 1526 LS, 1576 Saxton
Thurlighe 1528 LP
Thyrlye 1535 VE
Thirlegh 1535 VE
Thirlye 1549 BM
Thurleigh al. *Raleigh* al. *Laleigh* 1641 BHRS viii. 174[1]

---

[1] The forms in the *Lincoln Registers* are worth quoting: *Lega* to 1471, except for *Leighe* (c. 1350) and *Thurleigh* (c. 1500). Then, *Lega* al. *Thyrlye* (c. 1490), *Lega* al. *Thirley* (c. 1500), *Thyrlegh*, *Thurly*, *Thirleigh*, *Thurleigh* (1514–70).

The etymology of this interesting name has already been given in BHRS viii. 174. It goes back to the dat. sg. *leage* of OE leah, 'clearing,' and would normally have developed to *Lee* or *Leigh*. From the common use of this word with the feminine of the definite article in such forms as OE *æt þære leage*, ME *at there lege* fresh forms arose. By a common process of colloquial misdivision of words, which is further illustrated in EPN *s.v.* æt, æsc, ac, we get forms *at the releye* and *at therleye* which yield the independent forms *Releye* and *Therleye*. This last has survived in the form *Thurleigh* and has preserved a trace of its earlier history in its unusual stressing of the last syllable. That is where the stress would normally fall in the expression *at there leye* from which it comes. The pronunciation of the final syllable shows the normal development stressed *leage* to Mod Eng [lai], found also in Leigh (Wo), Asthall Leigh (O) and in Sussex names such as *Ardingly* which also, though for some other reason, show the same stressing of the final syllable.

BACKNOE END

> *Bakenho* 1276 *Ass* (p), 1279 RH
> *Backeno* 1377 IpmR
> *Backnowe* 1535 VE
> *Bakenhoe* 1549 BM
> *Becknar Ho.* 1765 J

'Bac(c)a's hoh.' For this pers. name we may compare Backworth (Nb), Backwell (So), Bacton (He) and Baccamoor (D). In English records the name *Bacca* only occurs in LVD. It is best explained as an assimilated form of *Bad(e)ca*.

Turvey

TURVEY 84 E 4

> *Torueie, Torveia* 1086 DB, 1176 P (Chanc Roll), 1194 BM,
> 1287 *Ass*, 1315 Ipm
> *Turuea* 1168, 1170 P
> *Torfeia, Torfeye* 1176 P, 1302 FA
> *Turfeia, Turfeye* 1176 P, 1276 RH, *Ass*, 1287 *Ass*, 1293 Orig
> *Turueia, Turueie, Turuey(e)* 1175 P, 1198 Fees 10 *et passim*
> *Thorueye* 1247 *Ass*, 1249 FF

Thurueye 1254 FF, 1284 FA
Tourveye 1316 FA, 1347 Ipm

'Turf island,' one with good grass on it. *v.* eg. For the use of *turf* in p.n. cf. *turfhlawan* (BCS 537) and *turfhleo* (ib. 1201), though in the latter the reference is probably to 'sods' rather than 'turf,' thus 'shelter made of sods.'

LADYWELL, LEADWELL (local)

Landimareswell 1279 RH

This local survival is recorded in VCH iii. 114, but it is difficult to reconcile the earlier and later forms. *Landimareswell* is good ME for 'land-boundary's spring' and can hardly be a corruption for 'Lady Mary's well' as is apparently suggested in the VCH.

NORTHEY (Fm)

Northo 1242 Fees 885 (p), 1276 *Ass*

'The north hoh.' It is at the extreme north end of the parish.

PRIORY (Fm)

The Priory of St Neots held land here in 1279 (RH ii. 333).

TURVEY BRIDGE

pontem de Turueya 1276 *Ass*

WESTFIELDS BARN

Westfeld 1366 Cl

Self-explanatory. It is on the western edge of the parish.

Wymington

WYMINGTON [wimiŋtən] 84 A 5

Wimenton(e) 1086 DB, 1163, 1165, 1176 P (Chanc Roll) *et passim*
Wimetone 1086 DB, 13th *Dunst*
Wimunton 1169, 1175 P
Wimuncton 1174 P
Wiminton 1176 P (Chanc Roll)
Wimedune 1185 Rot Dom

Widminton 1195 Cur(P), 1211 Abbr
Wimmenton 13th *Dunst*
Wyminton, Wiminton 1203 FF, 1227 Ass, 1231 Cl *et passim*
Wimminton 1208 FF
Wymington 1284, 1316 FA, 1348 Ipm, 1372 Cl, 1390–2 CS
Wymmington, Wimmington 1287 *Ass*, 1780–1830 Jury *passim*
Wemyngtone 1428 FA, 1455 AD vi, 1492 Ipm

The first element is probably the OE pers. name *Widmund* and the second is tun, but the actual compounding of the two elements is curious. We should have expected *Widmundes-tun*, but Wymondham (Nf, Lei) and Wymondley (Herts) show similar absence of genitival *s* in compounds from *mund*-names. For a similar development of an OE -*mund* name, cf. Almington (St), DB *Almentone*, for which we have early forms showing that the first element is certainly *Ealhmund*.

III. BARFORD HUNDRED

Bereford(e) 1086 DB, 1202, 1227, 1240 Ass, 1284, 1302, 1316, 1346, 1428 FA

For the etymology of this name *v.* Gt Barford *infra*. The exact meeting-place of the Hundred is unknown. The village, in which it must have lain, is at the centre of the arc on which lie the parishes forming the Hundred.

Great Barford

BARFORD, GREAT 84 E 10

Bereforde 1086 DB *et passim* with occasional
Berford as in 1257 FF
Berfford 1542 AD vi
Bareford 1545 AD vi

This p.n., with others of the same form, offers considerable difficulties. In solving it the places with which we are concerned are Barford (Nf, Nth, O, Wa, W), Barforth (Y), for all of which DB has *Bereford* and an unidentified *Bereford* (BCS 466), 12th cent. copy, with good forms. There is also a Barford (Sr), for which we have the form *bæran ford* (BCS 627), but as the ME forms are uniformly *Bereford*[1] one may suspect that this is a dialectal spelling for *e*. There is also a *Beran ford* at which Offa signs a grant of land. We have no means of identifying it, but it may well have been the Surrey place itself. For this name (or names) it is possible to suppose an OE pers. name *Bera* cognate with OGer *Bero*, but it would be dangerous to find that name in all the other Barfords as it is more than improbable that so rare a pers. name should so frequently be compounded with the one element ford. Professor Ekwall (*Anglia Beiblatt* xxxii, 259) would take the first element to be OE *bera*, 'bear,' and refers to the fact that the Romans drew upon Britain for their supply of bears. He refers in further support of this to Barbon (We), DB *Berebrune* which might contain the animal name, and to *beran del* (BCS 398) which is similarly possible, and to *beran heafod* (BCS 120), though the last is more doubtful as there is an alternative form *bearanheafod*. He might have added as evidence for the bear in England the statement in the *Gnomic Verses*, 29

Bera sceal on hæðe

eald and egesfull

'the bear shall be upon the heath, old and terrible,'

but on the whole we may suspect that these bears were to be found in the wilder parts of the country, whether mountain or forest, rather than in Bedfordshire, and there is the further difficulty that while there are plenty of examples of ford compounded with names of domesticated animals such as Oxford, Swinford, Horsforth, Gosforth, Enford there is little certain evidence for their being compounded with names of wild ones. We have none with fox, brock, boar, *eofor* (= boar) and only one or two with hart. If we leave the one, or possibly two, examples of *Beranford* out of the question, and assume that they may contain the pers. name *Bera*, we are left with a series of Barfords for which we have no evidence beyond the initial *Bere-* found in DB. Ekblom (PN W *s.n.*) suggests that this is OE *bere*, 'barley,' but 'barley-ford' does not yield any satisfactory meaning, first because it would be difficult to explain the absence of wheat, rye, bean and other crop-fords, and secondly

---

[1] *ex inf*. Mr A. Bonner and the Rev. H. R. Huband.

because with continually changing crops such a name would not be very useful. This suggestion may, however, put us on to the right track. It is clear from OE compounds like beretun, berewic, bern, from *bere-ærn*, 'barn,' *bere-flor*, 'barn-floor,' that in OE the word *bere* had, side by side with its specific sense, developed a general sense such that it could be applied to all cereals much in the same way as *corn* is at the present time. In those days of bad roads, when the means of communication often consisted in little more than muddy tracks and ill-tended fords, it may well have been that a road or ford which would carry a good load of corn might be distinguished as a *bere-* or corn-road or ford. Two old names suggest the possibility of such a compound. In an original charter of 697 (BCS 97) we have *bereueg* in a list of bounds and this can hardly be explained in any other way than as a compound of bere and weg. There is also the old Barlichway Hundred in Warwickshire, which certainly looks like a similar compound.

To sum up, while it is clear that for one (or possibly two) of the Barfords one must assume a lost pers. name *Bera* or (more doubtfully) the presence of a bear, it is probable that in most if not all of the others we have a name descriptive of a ford of sufficiently good bottom, natural or artificial, to carry a good load of corn.

Birchfield (Fm)

This is really a manorial name. Hugh de *Breteville* held a knight's fee in Great Barford in 1166 (RBE) and this afterwards became known as Brytvilles Manor (1521) and Birchfields (1692). *v*. VCH iii. 182. Cf. the similar development of Spinfield (Bucks) from *Espineville* (PN Bk 189).

The Creakers

> *Crewkers* 1539 NQ i
> *Crecors* 17th NQ i
> *Crakers* 1766 J

This is a manorial name. Hamo de *Crevequer* held two hides in Barford in 1242 (Fees 886), and the present name is a corruption of *Crevequer's*.

NORTHFIELD (Fm)

Campus del North 1227 FF

Self-explanatory. It is at the north-west corner of the parish.

Colmworth

COLMWORTH 84 C 9

Colmeworde, Culmeworde, Colmeborde 1086 DB
Colmwurda 1167 P
Colneworth 1202 Ass (p)
Culmw(u)rth 1202, 1240 FF
Colmew(u)rthe 1203 FF (p), 1227 Ass, 1273, 1276 BM
Comwrth, Comord 1219 FF
Colmewrdh 1220 LS
Cumewrth 1227 Ass
Colmorth 1227 Ass, 1247 *Ass*, 1262 FF, 1276 *Ass*, 1322, 1334,
    1347, 1414 BM, 1428 FA, 1492 Ipm, 1526 LS, 1549 Pat
Columwrth 1227 Ass
Culmwrth 1240 FF
Colewurth 1240 Ass
Colmeswurth 1247 *Ass*
Colmworth Hy 3 BM, 1247, 1276 *Ass*, 1333 Ch, 1373 Cl
Colmurþe c. 1272 BM
Colmsworth 1276 *Ass*
Colmeworth 1276 *Ass*, 1373 Cl
Colmorde 1284 FA, 1287 *Ass*, 1297 *SR*, 1302, 1316 FA, 1324
    Cl, 1330 Ipm, 1346 FA, 1387 Cl
Colmoord 1427 BHRS ii. 41
Colmorth al. *Colmord* 16th BHRS viii. 140
Coulemorth c. 1600 *Linc*

'Culma's enclosure' *v.* worþ. The name *Culm* is found once in
OE and is taken by Skeat to be the strong form of the name *Culma*
which lies behind this name. It may also be found in Culmington,
cf. Bowcock PN Sa 81. There is also an OE name *Cylm(a)*
which Skeat takes to be a mutated form of this name. Both alike
are difficult of explanation. No Germanic stem from which they
could have come is known. For *Cylm* there is a good deal to
be said in favour of interpreting it as from *Cynehelm* (cf. Mid-
dendorff 36), indeed as a hypocoristic form of that name, formed

G

by the same phonetic process whereby at a much later date we get Kilmiston (Ha), containing the pers. name *Cœnhelm*, and Kilmington (D) from the same name. If so we cannot take *Cylm* as a mutated form of *Culm*, and we should interpret the latter as a similar hypocoristic form for *Cuðhelm*, explaining *Culma* as a weak side-form of this name. The common ME spellings with *o* for *u* seem ultimately to have affected the pronunciation of the p.n.

COLLEYHILL

Collow Hill 1607 *Terr*

This form shows that this place is probably a *low* (*v.* hlaw) rather than a *ley*. Is this by any chance for *Colm-low*, marking the burial place of the one-time owner of Colmworth?

THE DEAN (6″)

Dene 1240 FF

Self-explanatory. *v.* denu.

LANGNOE (lost)

Langeho, Langehill 1197 FF
Langenho(u) 1276 *Ass*, 1297 *SR* (p), 1323 BM
Langenhoo (wood) 1387 Cl

OE *æt þæm langan hōhe*, 'at the long hoh.' The alternative use of *hyll* and *hoh* in 1197 is interesting.

Eaton Socon

EATON SOCON 84 C 11

Etone 1086 DB, 1199 (1253) Ch, 1220 LS *et passim*, c. 1360
    Linc (*juxta St Neots*)
Etton 1202, 1227 Ass
Eaton 1208 BM, 1247 *Ass*
Soka de Eton 1247 *Ass*
Etton Beauchamp 1276 *Ass*
Eytone 1428 FA
Eaton cum Soca 1645 NQ iii

'River-farm' (*v.* ea, tun). 'Eaton Socon, as its name implies, was a soke or liberty in the 13th cent., which was free of suit at the Hundred Court of Bedford' (VCH iii. 199). The distinctive epithet does not seem to have been applied until the 17th cent. when the form *Socon* (rather than *soke*), from OE *sōcn*, ultimately came to be adopted. The *Beauchamp* family held a manor in Eaton.

BASMEAD MANOR

Bassemeye 1271 Coroner
Basmey(e) 1328, 1386 Cl

This manor is reputed to take its name from the family of *Baa* or *Bathonia*, i.e. Bath, who held land in this parish as early as the reign of John (VCH iii. 192). The early forms are inconsistent with this and suggest that the first element is the OE pers. name *Bassa*. What the second is cannot be determined. The present *mead*-suffix seems to be due to the influence of the neighbouring Bush*mead*.

BEGGARY

la Beggerie 1227 Ass, 1276 *Ass* (p)
(la) Beger(y) 1241 FF, 1247, 1276 *Ass*, 1372 Cl
(la) Begerie 1247, 1287 *Ass*, 1351 BM
le Beggerye 1374 IpmR
Begwary 1656 NQ iii

The word *beggar* (ME *beggere, beggare*) is not unknown in English place-names for it is found in Beggearn Huish (So) and it has a parallel in the frequent use of OE *loddere*, 'beggar,' in p.n. Cf. *loddere beorg* (BCS 1047), *loddra well* (ib. 1282), *loddera wyllon* (ib. 887), *lodder þorn* (KCD 796), *loddere lake* (BCS 34), *lodderes sæccing* (ib. 491), *lodres wei* (KCD 1367), *loddera-stræt* (BCS 895). The first element in this name may therefore be taken to be ME *beggere*, 'beggar.' The situation of Beggary forbids our taking the second element as OE eg, 'island,' even in the extended sense in which it was used in earlier times, and the uniform final *-ie, -ye* to the exclusion of the more common *-eye, -eie* is also against this view. Rather we must take the suffix to be that in ME *baronie*, 'the domain of a baron,' for which the NED gives a quotation from Robert of Gloucester (1297). Presumably a *beggerie* is the 'domain of a beggar,' descriptive of land so poor that its tenants must always be beggars. This seems more probable than to assume that the 17th cent. *beggary*, used of a place which beggars haunt and derived from *beggary* in its earlier sense of 'action of begging,' was already in use in the 13th cent. For such terms of abuse, cf. Weekley's note in MLR xvii. 412. A good early example of this type of name may be derived from the story told by Henry, Bishop of Winchester, that when he was Abbot of Glastonbury (c. 1125) a certain knight deceived him as to the value of a piece of land he held of the abbey by giving it the name of *Nullius proficui*, i.e. 'of no value.'

BUSHMEAD

> *Bissop(es)med* 1227 Ass, 1231 FF
> *Bissemedwe, Byssemedwe* 1227 Ass, 1276, 1287 *Ass*
> *Bissemede* 1227 Ass, 1236 BM, 1240 FF, Ass, 1243 Ipm
> *Bissemade* 1227 Ass, Hy 3 BM, 1242 Fees 886
> *Bissopemedwe* 1234 FF
> *Bismede, Bysmede* 1236 FF, 1276 *Ass*, 1286 Dunst
> *Bissepmede* 1239 FF
> *Biscopemed* 1247 *Ass*
> *Bysmade, Bismade* 1276 *Ass*, c. 1750 Bowen
> *Bishmedue* 1310 Cl
> *Bysshemede, Bisshemede* 1315 Ipm, 1387 Cl, 1523 BM
> *Bysshemade, Bissh(e)made* 1382 Cl, 1389 IpmR, 1390–2 CS, 14th Gest St Alb
> *Bush(e)mede* 1399 IpmR, 1427 BHRS ii. 41, 1526 LS

This is clearly 'Bishop's mead' (*v.* mæd), but no bishop is known to have had anything to do with it. On the other hand, in 1309 (Pat) we have mention of one Hugh *Bisshope* of Eaton Socon who was accused of trespassing in Eaton and Cadbury. Cadbury (*v. infra*) was at the Bushmead end of Eaton parish and it is just possible that the *Bishop* family were already here at the beginning of the 13th cent. and gave their name to Bushmead[1].

CADBURY MANOR (lost)

> *Kadberia, Cadebyri* Hy 2 (Hy 3) *St Neot* 81, 1208 BM
> *Cad(d)esbiry* 1278 QW
> *Cadesbury* 1287 *Ass*

---

[1] For this note we are indebted to Miss E. G. Withycombe.

Catbury, Catburie 1309 Pat, 1377 Cl, 1518 VCH iii. 197
Catebury 1314 Abbr
Cadebury, Cadbury 1325 Ipm, 1380 IpmR, 1382 Cl

This is presumably from OE *Cadan-byrig*, 'the burh of Cada,' a pers. name found also in Caddington *infra* 145 and probably of British origin; the only difficulty about this is that it leaves unexplained the three forms with a *t*. If we start with initial *Cat-*, the *d* forms could be explained as due to voicing before the *b*, but the only pers. name which would fit is the ON name *Kati*, and a compound of *byrig* with such a Scandinavian pers. name is highly improbable. It should be added that *Cate's* Wood in this parish is supposed to contain a last trace of the old manorial name (cf. VCH iii. 197)[1].

DULOE

Diuelho, Dyuelho 1167 P, 1227 Ass, 1228 BM, 1247, 1287 *Ass*, 1297 *SR*
Duuelho n.d. *St Neot*
Deuelho 1227 Ass, 1512 AD vi

This is a difficult name. Professor Ekwall suggests that there may have been an OE **dyfel*, 'peg,' cognate with Ger *döbel, tübel* (from **dubila*), 'peg,' cf. *dowel*, used of a peg in NED. The compound might then be descriptive of a hoh which, by its shape, suggested a peg or plug. Topographically this is not impossible.

There is a nickname *Deule*[2], presumably from OE *dēofol*, found in the Beds DB in the name of *Alwin Deule*, a tenant in Pertenhall, Tempsford, Clifton and Chicksand. If he or some-one else so nicknamed had given his name to the place we should have expected *Deulesho*. One must therefore probably reject the suggestion made in VCH i. 297 *n.*, that his name might be found in Duloe. With more certainty one may reject association with the Deyville family, cf. VCH iii. 190.

EATONFORD (Fm)

Forda 1297 *SR*

Self-explanatory.

[1] The only map-record that has been noticed is *Cadbury* Lane (1766, J), the name of the lane which runs E. and W. to Upper Staploe.
[2] Cf. the pers. name *Deuleward* (13th cent.) in Thurgarton Cartulary 115.

GARDEN WOOD

Gardynesgrave 1387 Cl

This may be a compound of the word *garden*, but 'grove of the garden' does not give much sense. Alternatively the proper form may be *Gardeynesgrave* from ME *gardeyn*, 'guardian, warden,' also a 'justice of the peace.'

GOODWICK

Godynewyk E 1 BM
Godewyk 1276 RH, *Ass*
Godywyk 1297 *SR*

This is probably from OE *Gōdgife-wīc*, 'the wic or dairy-farm of a woman named *Gōdgifu*' (Latinised *Godiva*). This explanation involves the assumption that the first form is an error of transcription for *Godyuewyk*, but that is no difficulty as compared with the alternative that the first element is the pers. name *Godwine*. Even if we allow for early loss of medial *w* it is a good deal more difficult to see how *Godynewyk* could become Goodwick than to accept the alternative proposed above.

HONEYDON

Huneden 1227 Ass
Honyden, Honiden 1240 Ass (p), 1247 *Ass*, E 1 BM, 1276, 1287 *Ass*, 1297 *SR*, 1428 IpmR, 1433 BM, 1535 VE
Huniden, Hunyden 1247, 1287 *Ass*
Honydon 1276 *Ass*, 1330 Ipm

'Honey-valley,' presumably from its plentiful honey. *v.* hunig, denu.

STAPLOE

Stapelho(u) 1227 Ass, 1228 FF, BM, 1247, 1276 *Ass*, 1292 Ipm, 1318, 1380 Cl
Stapilho, Stapylho 1276 *Ass*, 1512 AD vi
Stapulho 1297 *SR*, 1330 Ipm, 1387 AD vi
Staplo 1512 AD vi
Staplowe 1549 AD vi
Staplehow 1619 BM

'Hill once marked by some post or pillar' *v.* stapol, hoh. Cf. *Stapelho* (1292) in King's Ripton (*Hunts Ct Rolls*), also Duloe *supra* 57.

SUDBURY (lost)

> *Subberie* 1086 DB
> *Sutburn* 1185 P
> *Sutbir'* 1236 FF
> *Suthbur'* 1242 Fees 867
> *Sudbyr'* 1276 *Ass*
> *Sudbury* 1315 Ipm

v. suð, burh and cf. Sudbury (Sf).

WYBOSTON [waibəsən]

> *Wiboldestone* 1080 DB
> *Wibaudeston* 1208 BM
> *Wy-, Wibaldestone* 1220 LS, 1227 Ass
> *Wiboldeston* 1227 Ass

The forms are just what one would expect, except for such as

> *Wilboldeston* 1276 *Ass*

until we get

> *Wyboston* 1297 *SR*, 1302 FA
> *Wybolston* 1488 Ipm, 1512 AD vi
> *Wiberson* c. 1750 Bowen

though the longer forms persist till the middle of the 15th cent.

OE *Wigbealdestūn*, 'farm of Wigbeald.' This common name enters also into Willington (Berks), Wobaston (St) and Whillington (He). The form *Wilboldeston* found once in the 13th cent. must be rejected as an error, for starting from an OE pers. name *Wilbeald* it would be impossible to explain the long vowel in the first syllable of Wyboston.

Goldington

GOLDINGTON 84 F 8

> *Goldentone, Coldentone* 1086 DB
> *Goldinton* 1167 P, 1199 FF, 1227 Ass *et passim*
> *Goldington* 1225 FF (p), Hy 3 Ch *et passim*
> *Goutinton, Goudinton, Goudington* 1227 Ass, 1247 *Ass*

'Golda's farm' *v.* ingtun. The pers. name *Golda* only occurs as the name of a moneyer before 1066 and the stem *gold* itself is hardly found in compound names before this date. There is therefore a presumption that Goldington is a name of late OE origin. It is remarkable that the best evidence for the early use of this stem in OE comes from certain place-names such as Gillingham (Nf), containing a pers. name *Gylda*, and Gillingham (K, Do) containing an assimilated form *gylla-* of this name. *Gylda* and *Gylla* are best explained as derived by mutation from a base *guld* (= gold).

NEWNHAM PRIORY

> *Neweham* 1198 FF, 1202 Ass, 1230 Cl, 1247 FF, 1276 *Ass*
> *Neuham* 1202 Ass, 1207 FF, 1242 Fees 886, 1283 Ipm, 1286 BM
> *Newenham* 1214 FF, 1218–27 BM, 1242, 1244 FF, 1247, 1276 *Ass*, 1285 Orig, 1287 *Ass*
> *Newham* 1221 BM
> *Nywenham* 1276 *Ass*
> *Newenham without Bedford* 1285 Fine

After this the forms with inflectional *en* or *n* are much the commonest.

Self-explanatory. *v.* niwe, ham.

PUTNOE (Fm)

> *Puttanho* c. 1053 (c. 1250) KCD 920
> *Putenehou* 1086 DB
> *Puteho* 1163 P, 1276 *Ass*
> *Puttenho* 1167 P, 1247 *Ass*, Hy 3 (1317) Ch, 1276 *Ass*, 1346, 1428 FA
> *Putho* R i (1286) Ch
> *Putenho* 1242 Fees 886, 1252 Ch
> *Potenho(u)* 1276 *Ass*
> *Pottenho(e)* 1276 *Ass*, Eliz ChancP
> *Poutenho* 1276 *Ass*, 1302 FA
> *Potenhowe* 1291 Tax

'Putta's hoh.' *Putta* is a well-established OE name, occurring again in Puttenham (Herts).

RISINGHOE (Castle, site of)

> *Risingeho* c. 1150 (c. 1230) *Warden* 84
> *Rysingho* 1252 Ch, 1331 QW
> *Risinho* 1286 *For*

Either OE *hrīsinga*-hoh, 'the hoh or heel of land of the *hrīsingas* or dwellers in the *hrīs* or brushwood,' or, alternatively, OE *Rīsinga-hōh*, 'the hoh of *Rīsa*'s people' (cf. Riseley *supra* 18). Hurstingstone (Hunts) *infra* 203 is a parallel formation to the first of these alternatives.

Ravensden

RAVENSDEN [ra·nzdən] 84 D/E 8

Rauenesden(e) c. 1150 (c. 1230) *Warden* 51 *et passim* to 1493
Raueneden 1225 FF
Ravesden 1284 FA
Ravensdene 1302 FA
Ravenysdene 1428 FA
Rounesden al. *Ravenesden* 1466 IpmR
Raunston 1528 LP
Ravensdell al. *Ravensden* 1631 BHRS ii. 105

'The valley of *Hræfn*.' The name is not on record in OE but its existence is made probable by such names as Raveningham (Nf), which we must assume to be pre-Scandinavian. In any case, the existence of an OE pers. name *(H)ræfn* is made highly probable by the appearance of a compound *Wlfreuen*, borne by a burgess of Wallingford in the 12th cent. The initial *w* of this compound, and the first *e* of its second element show that it is an English and not a Scandinavian name. Such a compound would readily give rise to a short form *(H)ræfn*.

MOWSBURY HILL

Morsebury 16th VCH iii. 212

This is for *Morinesbury*, the manor taking its name from the family of Ralf *Morin*, who held land here in 7 Ric. i (P).

TILWICK (Fm)

Tolewic 13th (15th) *Newn* 63
Tolewyk 1258 FF
Tylwyck Eliz ChancP

The early forms of this name can hardly be reconciled with one another and interpretation at present is impossible. The early medieval names *Tole*, *Tola* are of Scandinavian origin, but the change of vowel from *o* to *i* would be difficult to explain.

TRAYLESFIELDS

This is the last relic of the manor of Trayles (1493 Ipm), which takes its name from the Trailly family, who were holding here in 1272 (Ipm) and were lords of Yelden (the head of their fee) and of a manor in Willey Hundred in 1086.

Renhold

RENHOLD [renəld] 84 E 9

  Ramhale (sic) R 1 (1286) Ch
  Ranhale 1220 LS
  Ranehala Hy 3 (1317) Ch
  Ranhall 1227 FF
  Ronhale 1227 Ass, 1247 *Ass et passim* to 1428 FA
  Renhal(l) c. 1230 WellsL, 1276 *Ass*, 1434 IpmR
  Runhale 1247, 1276 *Ass*
  Ronhall 1274 Ipm
  Ronale 1276 RH, *Ass*, 1329 Ch
  Runnehale 1287 *Ass*
  Rondhale 1287 *Ass*
  Rounhale 1338 Ipm
  Ronehall 1498 Ipm, 1535 VE
  Ronnall 1526 LS
  Reynold, Raynold 1549 Pat, 1576 Saxton, 1605 NQ i
  Ronhall al. Runhall 1549 Pat, Eliz ChancP
  Ronhall, Reynold, Raynald 17th BHRS viii. 157[1]

Skeat may have got on to the track of the history of this very difficult name when he suggests that the first element is *Hranig*, a name found in the late-OE period. He was troubled by the final *g*, but Björkman has shown (NP 69) that this is simply an anglicised form of the Scand. name *Hrani*. This name it may be added seems to be found also in Ranby (Nt), DB *Rane(s)bi*. One would have liked to have more forms with inflexional *e* between the *n* and the *h*, but it is possible that the corrupt *Ramhale* is for *Ranehale* or even *Ranihale* and conceals at least one such. That this pers. name, even though it has a short vowel in ON, might

---

[1] The forms of this name in the *Lincoln Registers* run: *Ronhale* (1347–98), *Ronehale* (c. 1425), *Renhalle* (c. 1438), *Ronhale* (1452–94), *Renhold* (c. 1500), *Renhold* (c. 1550), *Reanold* (c. 1600).

appear as *Rone* in ME with an *o* is suggested by the name *Roni*, given by Florence of Worcester (12th cent. MS) as the name of an earl of Hereford in the 11th cent., which Skeat takes to be a later form of this name. This would explain the *Ran-* and *Ron-* forms (cf. forms of Sandy *infra* 107). The *Run-* forms we must then take to be inverted spellings for *Ron-*, due to the very frequent replacement of *u* by *o* before *n* in ME spelling. No explanation of the *Ren-* forms or the *Roun-* form can be offered. They cannot be related to the others and show a degree of irrational variation which is seldom paralleled. The suffix is clearly from OE healh. This developed to *-hale* and (as often) was perverted to *-hall* and then, by a common vulgarism, fully illustrated in Wyld's *Colloquial English* 309, a final *d* was added.

ABBEY (Fm)

This was held at one time by Newnham Priory (VCH iii. 215).

BROOK FARM

cf. *Brocfurlong* 1227 FF

Self-explanatory.

GREAT EARLY GROVE [a·li]

Arnele 13th (15th) *Newn* 118, 1297 *SR* (p)

'The clearing of *Earna*' or 'of the eagles.' *v.* earn, leah.

GADSEY BROOK (6")

Godeshoslade 1239 (15th) *Newn* 63
Goteshoslade c. 1300 (15th) *Newn* 95

'The slæd by God's hoh,' *God* being a pet form of one of the OE pers. names in *God-* and not the name of the divinity. The change of vowel is probably due to the same unrounding process (combined with a shyness in using the name of God) which has given us 'by Gad.' Cf. Wyld, *Colloquial English*, 240.

HOWBURY (Hall)

Elizabeth, wife of John de Horbury, held land in Renhold in 1276 (RH i. 2) and the manor is called Hobury in 1549 (Pat). It is impossible to be sure whether the name is a manorial one, the family having come from some other place of that name (the only one that has been noted is one in Yorkshire) or if *Horbury* is of genuine local origin and comes from horh and burh when it would mean 'dirty bury.'

SALPH END [sa·f end]

> *Salcho(u)*   1086 DB, 1242 Fees 886, 1247 *Ass*, 1291 Tax, 1428 FA
> *Salhho* 1227 Ass (p)
> *Salho* 1247 *Ass*, 1276 *Ass* (p)
> *Saluho* 1276 *Ass*, 1287 *Ass* (p), 1302 FA
> *Salugho* 1287 *Ass* (p)
> *Saliho* 1297 SR
> *Salfho* n.d. (15th) *Newn* 109
> *Salpho* 1377 IpmR
> *Salphobury* 1535 VE
> *Safe End* 1766 J

'Spur of land with willows on it[1]' *v.* sealh, hoh. For the pronunciation we may compare Saughtree (Sc) pronounced [sæftri·].

Roxton

ROXTON 84 D 11

> *Rochestone, Rochesdone* 1086 DB
> *Rokesdun* 1220 LS, 1247 *Ass*
> *Rokesdon* 1227 Ass, 1232 FF *et passim* to 1500
> *Rokesden* 1232 Cl, 1241 FF *et passim* to 1300
> *Rokisdun* 1235 Ch
> *Rockesden* 1287 *Ass*
> *Rockesdon, Rokkesdon* 1297 SR, 1330, 1338 Cl
> *Rokeston* 1398 IpmR
> *Roxton* al. *Rokesdon* 1449 IpmR

This is probably '*Hrōc*'s hill' rather than 'rook's hill.' The existence of such a name, derived from the bird-name, is certain (cf. Mawer, *Animal and Personal Names* in MLR xiv. 241) and it is hardly likely that a hill would be named from a single rook. *v.* dun.

---

[1] There is a considerable number of willow stumps left upon the banks of the brook at Salph End (*ex inf.* the Rev. F. W. Breed).

CHAWSTON [tʃɔ·sən]

Chauelestorne, Calnestorne 1086 DB
Caluesterne 1167 P, 1203 Cur
Chaluesthorn 1180 P (p)
Chau(u)esterne 1202 FF, Ass
Calsterne 1220 LS
Chaluesterne 1227 Ass, 1239 FF, 1242 Fees 867 *et passim* to
  1400
Chalueston 1227 Ass, 1302 BM, FA
Chalston 1418, 1478 IpmR
Chalsterne 1428 FA, 1441 BM
Chauston 1535 VE
Chawson 1826 B

'Cealf's thornbush,' as suggested by Skeat. Evidence for the
pers. name *Cealf*, derived from the animal-name, is given in the
article just referred to (p. 237). That we have a pers. name rather
than the animal-name is suggested by the mention in the
Newnham Cartulary 186 of another of his possessions, viz.
Chaluescrofte. For the sound-development cf. Chauson (Wo)
from *Cealfes-tun.*

COLESDEN

Colesden(e) 1195 P (p), 1202 Ass, 1220 LS, 1227 Ass, 1276,
  1287 *Ass et passim*
Cole(n)dene 1227 Ass
Collesden 1236 FF, 1276, 1287 *Ass*, 1338 Cl, 1415 Ipm,
  1428 FA, 1440 IpmR, 1441 BM, 1493 Ipm, 1535 VE
Colisden 1276 RH
Collesdon 1343 Cl
Colsdene 1346 FA
Collisden 1386 Cl
Coldsden 1592 AD vi

'Col's valley.' The evidence for an OE pers. name *Col* is set
forth *s.n.* Coleshill (PN Bk 227).

WOODEND LANE

Wodende 1287 *Ass* (p), 1297 *SR*

Self-explanatory.

Wilden

WILDEN [wildən] 84 D 9

Wilden(e) 1086 DB, 1163 P, 1197 FF *et passim*
Wileden 1167 P
Willeden(e) 1167 P (Chanc Roll), 1346 Ipm, 1526 LS
Wildon 1194 P *et passim*
Wildana 1208 FF
Wyleden 1234 Cl, 1287 *Ass*
Wyledon 1264 BHRS v. 232
Weledene 1284 FA
Wyl(l)eden 1287 *Ass*
Willden 1287 *Ass*
Wyldene 1297 *SR et passim*
Wilden al. *Wileden* 1323 Abbr
Wyliden 1346 Ipm
Wilding 1798, 1800, 1820 Jury

'Valley of Wil(l)a.' *Willa* is a well-established OE name and the existence of an alternative form *Wila* is made probable by the patronymic *Wilinc*. Cf. Willington *infra* 99. *v.* denu. The *-ing* forms show the common vulgarism heard in [kitʃiŋ] for *kitchen*.

BROOK FARM

cf. *Brocforland* 1227 FF
Self-explanatory.

EAST END

Estende 1269 BHRS viii. 244 (p)

SEVICK END

No explanation of this name can be offered. It should probably be associated with the John de *Sebyok* who is mentioned in the 1287 Assize Roll in connexion with Beggary, four miles off.

IV. REDBORNSTOKE HUNDRED

Radeburnesoca, Radebernestoch, Radborgestoc, Ratborgestou, Ratborgestoche, Ratberbestoche 1086 DB
Reiburgestoch 1156 P
Redburnstowe, Redburnestou 1175, 1176 P
Redburnestoke 1183 P, 1202, 1227 Ass, 1428 FA

Redburgestok 1193 P
Redeburnestoke 1284 FA
Redburghstoke 1284, 1346 FA
Redbournestoke 1316 FA

It is much to be regretted that we know nothing of the meeting-place of this Hundred, for the name raises great difficulties. The suffix is either stoc or stocc, the reference in the latter case being to the 'tree-stump' at which the meeting took place. There has however clearly been confusion with the suffix stow, a confusion which has its parallel in the case of the lost name *Laverkestoke* in Bk (PN Bk 241 note). Confusion of this kind is not uncommon in long hundredal names, cf. Wixamtree *infra* 87. The confusion in the first element of the name is more baffling. On the whole it would seem best to take it as OE *Rǣdburh*, a feminine name recorded in LVD and surviving into the 12th cent. If that is correct, *Rædburgestoc* is one of the few hundredal names in which a woman's name forms the first element. From this point of view it may be compared with the Gloucestershire hundredal name which appears in DB as *Celfledetorne, Celfleode, Ceolflede* Hundred, which preserves the OE feminine compound *Cēolflǣd*. If we accept *Rǣdburgestocc*, 'Rædburh's tree-stump,' as being the origin of the name, the *-burn* forms must be interpreted as showing the common confusion of the suffixes burh and burna which is fully illustrated in PN NbDu 270, and would here be facilitated by false association of the name with 'red' and 'bourne.'

Ampthill

AMPTHILL [æmtəl] 84 J 7

Ammetelle 1086 DB
Amethull(e) 1202 Ass (p), 1227 Ass, 1242 AD vi, Ch, 1276
    Ass, 1305 Ipm, 1316 FA, 1323 Cl
Aunthille c. 1230 WellsL
Hamethull 1242 Fees 890
Amett(e)hull 1247, 1276 *Ass*
Ammethull 1247 *Ass*
Ametull(e) 1302 FA
Ampt(e)hull(e) 1276 *Ass*, 1346 FA, 1350 Ipm, 1362 Cl, 1390-
    2 CS, 1428 FA, 1452 AD vi

Amt(e)hull(e) 1276, 1287 *Ass*
Hampthull 1344, 1366 Cl
Amptyll(e) 1429 Chron St Alb, 1509 AD v
Amptehill 1442 AD ii
Anthill, Antyll 1528 LP, 1530 NQ i, 1675 Ogilby
Amptell, Amptle 1535 VE, 1636 HMC iv, 1675 Ogilby

OE *æmette-hyll*, 'ant-hill,' presumably, as Skeat says, 'a some-what jocular appellation,' though it might be that the site was simply ant-infested. Cf. *Amethulle* (Hy 3) as a field-name in Arlesey (*Proc. Soc. Antiq.* (2nd Ser) iii. 307).

DOLITTLE MILL (6″)
'The first mill on a stream is often known as Doolittle Mill' (BHRS iii. 219).

Cranfield

CRANFIELD [kræmfiˑld] 84 H 5
Cranfeldinga dic 969 BCS 1229
Crangfeld 1060 (14th) KCD 809, 1253 BM, 1077 (17th) Chron Rams
Cranfelle 1086 DB
Cranefeld c. 1125 (c. 1350) Rams, 1227 Ass, 1228 FF, 1287 QW, 1302 FA
Craunfeld c. 1125 (c. 1350) Rams, 1227 Ass, 1247 *Ass*, 1251 Ch, 1276, 1287 *Ass*
Crancfeld 1133–60 (c. 1350) Rams, c. 1300 Chron Rams
Cramfeld 1202 FF, 1287 QW
Cramfelt 13th AD iii
Cranfeld 1202 Ass, 1220 LS, 1227 Ass, 1247 *Ass*
Cranefeud 1227 Ass, 1242 Fees 869
Craumfeld 1287 QW, 1311, 1388 Cl, 1390–2 CS
Crang(e)feud 1293 AD i
Crainfeld 1526 LS

'Open country frequented by cranes' *v.* **cran**, **feld**, apparently with an alternative form in which **cranoc**, the other OE name of this bird, was used. This would explain the *Crang-* and *Cranc-* forms.

BOURNE END

la Burne 1227 Ass (p), Hy 3, 1293 AD i
le Burnehende 1306, 1312 *Hunts Ct Rolls* (PRO)

'The quarter of Cranfield which lies by the **burna** or stream.'

GROVE (Fm)

Brechegrove 1294 AD i

Self-explanatory except for the first element in the early form. This is the dialectal *breach*, 'land broken up by the plough' (ME *breche*), discussed in PN Bk 60.

LEYS (Fm)

Lees Boscus c. 1250 (c. 1350) Rams

This is the plural of ME *leye* from leah and denotes 'clearings in woodland.'

PERRYHILL

cf. *Piriecroft* 13th AD iii
Pyriefeld 1300 *Hunts Ct Rolls* (PRO)
Pyrebroke 1312 *Hunts Ct Rolls* (PRO)
del Pyrye 1317 ib. (p)

v. pirige. All these places must have taken their name from some prominent pear-tree. Cf. Perry (Hunts) *infra* 271.

WHARLEY END

Horle 1244 (c. 1350) Rams, 1306 *Hunts Ct Rolls* (PRO)
Wallend or *Warle End* 1766 J
Warlend 1853 O
Walley Farm 1853 O

'Dirty clearing' *v.* horh, leah. For the development of initial *h* to *w* or *wh*, cf. *Worrage*, a colloquial form of *Horridge* for Hawridge (Bk). *a* for *o* shows a South Midland dialectal development of *or* to *ar*, cf. EDG 76.

WOOD END

Wodende c. 1250 (c. 1350) *Rams*
le Wodehende 1306 *Hunts Ct Rolls* (PRO)

Self-explanatory.

H

Elstow

ELSTOW [elstǝ] 84 G 8

Elnestou 1086 DB, 1174–81 (13th) *Dunst*
Elvestou(e) c. 1150 Reg Dun, 1174–81 (13th) *Dunst*
Alnesto 1177 BM
Elnesto(u) 1177, 1203 BM
Alnestow 1182 P, 1194 Cur(P), 1202 Ass, 1214 Abbr, 1220
    LS, 1227 Ass, 1247, 1276, 1287 *Ass*
Auuestowe (sic) 1197 FF, 1247 *Ass*
Aunestow 1202 Ass, 1219 FF, 1247, 1276, 1287 *Ass*, 1286
    Dunst, 1315 Ch
Elnestowe 1232 Pat, 1242 Fees 868, 1276, 1287 *Ass*, 1291 NI,
    1305, 1310 Cl, 1316, 1346 FA, 1389 Cl, 1510 LP
Aluestowe 1239 FF, 1247 *Ass*
Eluestowe 1247 FF, *Ass*, 1264 FF, 1289 Ipm
Elenestowe 1258 Pat, 1526 LS, 1530 LP
Helenstoe c. 1270 Gerv
Anestowe 1276 *Ass*
Eylenestowe 1287 *Ass*
Eleynestowe 1415 BHRS ii. 35
Elnystowe 1428 FA
Elmestowe 1518 *Award*
Ellenstowe 1518 *Award*
Elvestowe al. *Elstowe* 1589 BHRS iv. 14
Evelstow, Elstowe Eliz ChancP
Elvestow 1766 J

The forms in the *Lincoln Registers* are uniformly *Elnestowꞓ*
from 1300 to c. 1520, except for *Elstowe* (c. 1475). Then c. 155c
we have *Elnestowe* and (once) *Elenestowe*.

The first point to be made in regard to this difficult name is
that the first element must in OE of the first half of the 11th cent.
have begun with *Æln-*, for so only can we explain the *Aune-*
forms. The *Elue-* forms were, in the first instance, errors of
transcription, but in course of time tended to establish themselves as the true forms, possibly even in pronunciation (cf.
the inverse process in the replacement of the correct *Ioua* by
Iona). This being so we may rule out of consideration all connexion with St Helena (OE *Elene*), for it would not explain the

early forms, though it should be added at the same time that the actual dedication of the Chapel of St Helen as a parish church may have influenced some of the later spellings. How can we explain an OE *Ælnestowe* of c. 1050? Skeat rightly rejects the idea of taking this as from earlier *Æðelwines-* or *Ælfwines-stow*. The *w* in these names is not thus early lost (cf. IPN 173). He inclines to *Ælfnōþes-* or *Æþelnōþes-stow*, or possibly to the names *Æþelhūn-* or *Ælfhūn*, and adduces examples of the reduction of these names in various place-names, but in all these the DB or other early forms show approximately the full pers. name. The reduction of the name to anything like that supposed to be found in the DB form of Elstow belongs to a far later date. One can only relate an 11th cent. *Ælne* to such names if one believes it to be a definite OE hypocoristic form for one of them. Such forms, in which the blending of the two elements has taken place, are known in OE (IPN 173) and we might assume either an OE *Ælna* for *Ælfnōð* or, less probably, for *Æþelnōð*. If so, the full form of the name in OE would have been *Ælnan-stow*. Personal-names when compounded with stow seem as a rule to be those of saints, but there are exceptions, such as Alstoe (R), from OE *Ælfnoþesstow*.

Alternatively OE *Æll(e)n* might be a derivative of such a name as *Ælla* with an -*en* suffix, such as is found in certain OE names. Cf. IPN 171 n. 3 where mention is made of the probability of such names as *Cūþen, *Eaden. To these may be added **Æt(t)en* from *Ætti* which seems to lie behind Adstone (Nth), DB *Ateneston*, Northants Survey *Atteneston* and possibly *Peden in Pensham (Wo), *Pedneshamm* BCS 1282. Such names have been discussed at some length by Ritter 193 n. 2. *v.* Addenda.

MEDBURY

> *Meidebir'* 1227 Ass
> *Mayden(e)byr'*, *Maydenbur'* 1276 *Ass*
> *Maydeburn'* 1287 *Ass* (p)
> *Maide-burie* 1616 NQ iii

OE *mægða-byrig* (dat.) or *mægdenabyrig*, 'maid(en)s' fort' (*v.* burh), the exact equivalent of Germ. *Magdeburg*, but why these places were so called it is difficult to say. There are a good many modern place-names in which the element *Maiden* appears, combined with Way, Bower, Castle, and then it often refers to an ancient road or earthwork, but for none of these have we evidence of the early use of the name. In OE charters we only have *mægidna brycg* (BCS 428) and *mædena coua* (BCS 948), perhaps so called because frequented by maidens. Maidenhead (Berks) goes back to at least the 13th cent. and means 'maidens' landing-place,' which Skeat suggests may have been so called because there was an easy landing-place here. Similarly *Maidenesford* in a Beds fine of 1202 is presumably a very shallow ford in contrast to the *Mucheleford* mentioned immediately after. Maidencourt (Berks), earlier *Maidencote*, goes back to the 13th cent. and may well be 'dairymaids' cote.' Note also Maidwell, Maidford (Nth). Maiden Castle, applied to the Castle at Edinburgh (c. 1600), is found in Latin as *Castrum Puellarum* and, as suggested in the NED, may mean 'castle which even maidens could defend.' That is of course a possible sense for Medbury, but there is no evidence for any kind of 'castle' or even earthwork. Perhaps it may be 'manor' which was at one time held by 'maidens,' but these suggestions for this name and those for the others are more or less idle speculations. Such names have arisen from particular incidents or ideas which are now lost beyond recovery.

Flitwick

FLITWICK [flitik] 95 B 7

Flicteuuich(e) 1086 DB, 1175 P

Flittewik, Flittewic, Flyttewyk 1220 LS, 1227, 1240 Ass, 1255 KS, 1286 Dunst, 1316, 1346 FA, 1390–2 CS

Flitwik, Flitwyk, Flitwic, Flytwyc 1227, 1240 Ass, 1253 Ch, 1276 *Ass*, 1489 Ipm, 1509 AD v

Fletwyk(e) 1242 Fees 869, 1321, 1353 Ch, 1525 LP

Flittewyke 1247, 1287 *Ass*, 1296 Ipm

Flytewyk 1247 *Ass*

Flettewyke 1276, 1287 *Ass*, 1296, 1323 Ipm, 1398 Cl, 1452 AD vi

Fletewyk 1276 *Ass*, 1314 Ch, 1368 Cl

Flotewyk 1276 *Ass* (p)

Flythewick 1284 FA
Flutewyke 1297 *SR* (p)
Fletwyk 1321 Ch, 1353, 1368 Cl
Flitwyke 1428 FA
Fleetwick 16th BHRS viii. 149, c. 1640 *Linc*
cf. *Flithull* 1287 QW

v. wic. For this name, *v.* Flitton *infra* 148. It means 'dairy-farm on the stream.'

DENEL END

Dunhull(e) 13th *Dunst*, 1276 *Ass* (p), 1321 Ch, 1331 QW
Denhull 13th *Dunst*

This would seem to be a compound of dun and hyll. At first sight the compound seems redundant, but there is a good deal to be said for interpreting *dun* as open country, *down*-land at times, rather than as hill, and if so the compound describes open hill-country. One might suggest the adj. *dunn* as the first element, but there is no evidence for the topographical use of that term. In the one charter in which we have *dunnen cumb*, *dunnen dic* it is certain that it is a pers. name for these are in *Dunnestreatun* (BCS 229), which survives as Donnington.

The phonology is not easy. There is good precedent for *Dun-* becoming *Din-* in p.n. (cf. PN NbDu 258 and Dinton (Bk), possibly also Dinton (W)), but less evidence for that *Din-* becoming *Den-*. There are, however, two examples in Buckinghamshire, Denham in Quainton and the 16th cent. forms *Denton*, *Dennington* for Dinton (PN Bk 93, 159), and one in Northants, Denshanger, 1227 Ass *Duneshangre*. It is difficult to say how far this change represents a definite phonological development or may, alternatively, be due to a tendency to replace an unfamiliar initial *Din-* by the more common *Den-*.

EAST END

Eston 1321 Ch
Estowne 1519 AD vi

Originally 'east farm,' *v.* east, tun.

PRIESTLEY [prestli]

> *Prestelai* 1086 DB
> *Prestele(ye)* 1196 P, 1202 Ass *et passim* to c. 1350
> *Prestle(e)* 1321 Ch, 1373 Cl, 1403 IpmR

'Priests' clearing' *v.* leah. Whether it was a clerical endowment or what the relation of the priests to the clearing was we cannot now say. The normal development would have been to *Prestley*, and that can still be heard locally.

RUXOX

> *Rokeshoc* 1174–7 (13th) *Dunst*, 1286 Dunst
> *Rokeshac* 1174–7 (13th) *Dunst*, 1286 Dunst
> *Rokesac* 1220–45 (13th) *Dunst*, 1286 Dunst
> *Rokussoc* 1286 Dunst
> *Rokesekes* 1390–2 CS
> *Russoxes* 1535 VE
> *Roxax* 1657 NQ iii

Skeat is clearly right in taking this as from OE *Hrōces-āc*. The owner of this *ac* or oak bore the same name as the owner of Roxton *supra* 64. The later forms in *k(e)s* are pseudo-plural or pseudo-possessive and *ks* has become *x*.

Houghton Conquest

HOUGHTON CONQUEST 84 H/J 7

> *(H)oustone* 1086 DB
> *Octona* John (1227) Ch
> *Hocton* 1202 Ass, 1224 FF, 1247 *Ass*
> *Hoghton* 1220 LS, 1297 Ch, Cl
> *Hohton* 1227 Ass
> *Houton* 1242 Fees 868, 1274 Cl, 1276 *Ass*
> *Houcton* 1247 *Ass*
> *Houghton* 1287 *Ass*
> *Houghton Conquest* 1316 FA
> *Horton* 1675 Ogilby

v. hoh, tun. 'Farm on the spur of land.' The *Conquest* family are first associated with it in 1223 (FF).

BURY FARM

Houghton Conquest al. *Conquest Bury* 1549 Pat

This takes its name from the *bury* or manor (*v.* burh) of the Conquest family.

HILL FARM

Calewellehill 1224 FF

The form is probably for *Caldewellehill*. If the identification is correct the meaning is obvious.

REDDING'S WOOD

The last relic of Redding's Manor, held by Reading Abbey in 1242 (Fees 868).

THICKTHORN, GT and LITTLE (Fms)

Thykethornes 1276 *Ass* (p)

Self-explanatory. Cf. Thickthorn (Bk).

Kempston

KEMPSTON

Kemestan 1060 (14th) KCD 809, 1199 Abbr
Cœmbestun Edw Conf (c. 1350) Rams
Cæmbestune 1077 (17th) Chron Rams 202
Camestone 1086 DB, 1200 FF
Canbestuna 1124–8 Scott Hist Rev xiv. 372
Kembeston 1176 P (p), 1237 Cl, 1254 Orig, 1291 Tax, 1325 Ipm, 1402 BHRS i. 102
Kemeston 1189 P, 1199 FF, 1202 Ass, FF, 1220 LS, 1237 Cl, 1247, 1276 *Ass*, 1284 FA, 1287 *Ass*, Ipm, Cl, 1313 Cl, 1332 Pat
Kamistuna 1195 P
Camestun 1195 P
Chemiston 1199 P (p)
Cemeston 1199 Cur
Cambesdon, Kimbeston 1201 Cur (p)
Cambeston 1236 FF
Kemston 1236 FF, 1240 Cl, 1247, 1276 *Ass*, 1282 Cl, 1316 FA, 1334 Fine, Cl, 1346 FA, 1349 Ipm

Kembestun 1237 Cl, 1253 BM
Kemestun 1241 Cl
Kemyston 1242 Fees 868
Kempston 1247 *Ass*, 1290, 1355 Ipm *et passim*
Kempeston 1276 *Ass* (p), 1306 Cl
Kemmeston 1276 *Ass*
Camston 1328 Pat
Cameston 1332 Pat

Upon this name Professor Ekwall has contributed the following note:

The place is situated at a sharp bend of the Ouse. This suggests that the first element may be connected with British *cambo-* (Welsh *cam*) 'crooked.' There is in Wales a common place-name, which appears in slightly varying forms, as *Kemeys* (two on the Usk), *Cemmaes* (on Cemmaes Bay, Anglesey), *Cemais* (on the Dysynni) etc. The names in early sources appear as *Cemeis*, *Kemeis* etc. The places in question are situated on bays or at bends of a stream. No doubt the name *Cemeis* is in reality an old common noun meaning a bay or a bend of a river, and cognate or identical with Ir *camus*, Gael *camas* 'a bay' (see Owen's *Pembrokeshire*, i. 435). The same name is found in Cambois (Nb), Cams (Ha), IPN 26. The exact history of *Cemeis* offers some obscure points, but the word is evidently a derivative of *cambo-* 'crooked.' It presumably represents a form with an original *i* or *j* after the *s*, which owing to epenthesis came to form a diphthong with the vowel of the preceding syllable and which caused umlaut of the *a* of the first, that is an OWelsh *Cembes* with palatalised *s*, later *Cembeis*, *Cemeis*. At the time of the Saxon invasion the name would have its *b* left; we may compare *Cam Beck* (Cu), earlier *Camboc*, also a derivative of *cambo-* 'crooked.' The suggestion may be ventured that the bend in the Ouse on which Kempston stands, or a place near it had a pre-English name *Cembes* or the like and that from it was formed Kempston by the addition of *tūn*, just as Gloucester, Dorchester (O) were formed by the addition of *ceaster* to the British name. Some uncertainty prevails as to the relation between the umlauted form in Kempston and the non-umlauted one in Cambois, Cams, and between the variant forms of Kempston (æ, *a*, *e*).

Cambois, Cams may have been adopted earlier than *Cembes* in Kempston, before the British umlaut had operated. If so, we may assume that for the OW *e* in *Cembes* was substituted the umlaut of OE *a* before nasals, which appears in OE variously as *æ* and *e*. Both developments are recorded in Beds. But the first element of Kempston may have been adopted before British umlaut had taken place; if so, we must assume that an OE **Cambis* was substituted for the British form and this became OE *Cæmbes*, *Cembes* by Old English umlaut.

BARKDITCH (lost)

> *Barkesdig* 1200 FF
> *Barkedich* 1276 RH

Possibly from OE *beorca-dic*, 'ditch of the birchtrees,' *v.* beorc, dic, or a pers. name *Bark* (cf. ODan *Barki*). This is commonly supposed to be the same as that now known as *King's Ditch*, and this in turn is supposed to be part of the fortifications of Bedford due to Edward the Elder (VCH iii. 1).

KEMPSTON HARDWICK

> *Hardwyke* 1276 *Ass* (p)
> *Herdewyk by Kemeston* 1334 Ipm
> *Kempston Hardewik* 1485 Ipm

v. heordewic.

THE HOO

> *le Hoo* 1460 AD iii

v. hoh.

KEMPSTON WOOD

> *attewode de Kemston* 1287 *Ass* (p)

Lidlington

LIDLINGTON 84 J 6

> *Litincletone* 1086 DB
> *Littlingeton* 1180 P (p)
> *Litlingeton* 1199, 1202 FF (p), 1202, 1227 Ass
> *Litlington, Lytlington* 1204 FF, 1227 Ass *et passim*
> *Letling(e)ton* 1204 FF

Lutlingeton 1220 LS
Littlington, Lyttlyngton 1228 FF, 1247 *Ass et passim* to c. 1300
Lutlinton, Lutlington, Lutlyngton 1262 FF (p), 1287 *Ass*, 1346
    FA, 1358 BM
Lutelington 1274 Cl
Lidlington 1780 Jury
Lidlington al. *Litlington* 1806 Lysons

OE *Lytelingatūn*, 'farm of Lytel's people.' *v.* ingatun in
EPN 42. The same pers. name is found in Littleworth, Lilling-
stone (Bk), Littleton (Mx) and Litlington (C). Voicing of *t*
to *d* before *l* is common in dialect and it is noteworthy that two
of the counties in which Wright gives examples are Essex and
Suffolk (EDG § 283).

BOUGHTON END

Bowden End 1637 VCH iii. 305

There is not enough material to do anything with.

COMBE PARK (lost)

boscus voc. Cumbes 1276 *Ass*
le Cumbes 1287 *Ass*

v. cumb. This is one of the few parishes in Beds in which
the ground admits of anything which might be called a *coombe.*

ESCHEAT [estʃi·t]

The history of this curious farm-name has been given us by
the Rev. A. D. M. Gowie, Vicar of Lidlington, on the authority
of its present owner, Mr H. Lines. 'The history of the for-
feiture goes back to the ownership of Dr Small, who died insane
and intestate. The property which was in the Manor of Bedford
escheated to Hastings, 9th Duke of Bedford.'

HOLT (Plantation)

Holtebussches, Holtebroke 1330 Ipm
Self-explanatory. *v.* holt.

THRUP END

Trop 1276 *Ass* (p)
Throp or *Thruppe End* 1637 VCH iii. 305

Thropp End 1766 J
Thorp End 1826 B, 1853 O

v. **þorp**. The word is clearly used here of an outlier of Lidlington.

Marston Moretaine

MARSTON MORETAINE [maˑsən mɔˑtən] 84 J 6

Mersctuninga (gen. pl.) 969 BCS 1229
Merestone 1086 DB
Merstone 1086 DB *et passim*, c. 1335 *Linc* (*juxta Bedeford*)
Mersh(e)ton 1287 *Ass*, 1300 Cl, 1316 FA, 1361 Cl, 1390 IpmR
Merston Morteyn 1383 Cl
Marson Eliz ChancP, 1662 Fuller
Marston Morton 1666 NQ i
Marston Mortine 1840 Jury

'Marsh-farm' *v.* **mersc, tun**. On the Morteyn family in Bedfordshire, see the article on that name in BHRS ix. 5 ff.

ASHBROOK

Aschebroc, Ashebroke 1287 *Ass*
Self-explanatory.

BEANCROFT (Fm)

cf. *Benhull* 1232 FF
Beans were evidently a common crop in Marston.

GREEN LANE'S CROSSING (6″)

Grenelanehull 14th (15th) *Newn* 138 *b*
Probably the same 'green lane' is referred to in these two names.

HUNGERHILL (Fm)

Hungerhull 14th (15th) *Newn* 137 *b*
'Hunger-hill,' a common term of reproach for barren ground, *v.* Weekley as under Beggary *supra* 55. Cf. *Hungerhyll* in Biggleswade (13th), *Hungirhill* in Flitwick (13th), *Hungerhul* in Ellington (1322).

MARSTON THRIFT

boscus del Frith de Merston 1287 *Ass*

v. fyrhþe. For the form, cf. Salem Thrift *supra* 30.

ROXHILL

Wrocheshola 1180 P (p)
Wrokeshala 1186 P (p)
Wrochishill 1195 FF(P)
Wrokeshull 1219 FF, 1242 Fees 892 (p), 1347 Cl, 1348 Ipm
Wrogxhulle 1220 LS
Wroxill 1240 Ass
Wroxhull 1240 Ass, 1247 *Ass*, 1253, 1258 FF, 1276, 1287 *Ass*,
    1313 Ipm, 1346 FA, 1393 IpmR, 1428 FA
Wroxhill 1247 *Ass*, 1492 Ipm
Roxwell Eliz ChancP

'Wrocc's hill.' The pers. name *Wrocc* is not on record in OE
but must be inferred from Wroxall (Wt), KCD 768 *Wrocces-
heale*, and (Wa), Wroxton (O), Wroxham (Nf), Wraxall (W),
earlier *Wrockeshale* (Ekblom, PN W). For this name *v.* Zachrisson in *Studier i Modern Språkvetenskap* ix. 124.

SHELTON

Es(s)eltone 1086 DB
Sheltune 1197 FF
Selton 1202 Ass
Shelton 1227, 1240 Ass
Schelton 1262 FF

This has the same history as Shelton *supra* 19.

WOOD END

le Wodende 14th (15th) *Newn* 139

Self-explanatory. The district is on the western side of the
parish.

Maulden

MAULDEN 95 A 8

Meldone 1086 DB, 1287 *Ass*
Meudon 1152–8 NLC
Maldon(e) 1163–79 BHRS i. 120, c. 1186–95 NLC, 1220 LS
    et passim

Maudon(a) 1179 P, 1219 FF, 1242 Fees 887
Mealdon 1180 P (p)
Meaudun, Meaudon 1195 Cur(P), 1238 FF
Malden(e) 1388 Cl, 1428 FA, 1568 BM
Maldoun 1518 *Award*
Maulden c. 1550 *Linc*

Skeat's explanation that it is a dun marked by a mæl or cross is no doubt correct. It is the same name as Maldon (Ess). Some of the early forms are curiously Gallicised, perhaps because the alien priory of St Faith of Longueville held land there.

BREACH

ate Brache 1307 *Ass* (p)

For this element *v.* Grove Farm *supra* 69 and bræc.

KING'S FARM

A thegn of King Edward, two sokemen of King Edward and (TRW) 'a certain king's Bailiff' held land in Maulden, and one or other of these holdings may have given rise to the name.

LIMBERSEY [limǝsi]

Limboldesheye R i (1285) Ch
Limbode(s)heie, Lymbodeshey(e) 1200, 1201 Cur, 1202 FF, 1227 Ass, 1286 *For*
Limbotesheye 1302 FA
Lymboteseye 1331 QW, 1346 FA
Lymbottysheye, Lymbottes Heye 1535 VE, 1549 Pat
Limersey 1766 J

'Linbeald's (or Lindbald's) enclosure' *v.* (ge)hæg. This name like *Breach* suggests that we are here in ancient woodland. Other *hays* in this parish were Eadmær's (1254 BHRS viii. 220 *Admereshey*, 1600 *Deed, Admersey Leas*) and Toli's (*Tholeshey* 1230 FF, *Tolyeshey* 13th AD iv). Limbold shows early assimilation of *nb* to *mb*, followed later by loss of *l* from the consonant-combination *lds*. *Lin* names are not common in OE or indeed in the Germanic languages generally. In OE we only have *Linbald* in DB and *Linxi* (DB) and *Lynsige*, the name of a late moneyer, both of which forms would seem to go back to OE *Linsige*. These names are probably of continental origin, cf. *Maðelpert* in Meppershall *infra* 171.

Millbrook

MILLBROOK   84 J 6

Melebroc 1086 DB, 1185 (c. 1200) *Templars*, 1285 Ipm
Molebroke R i (1286) Ch, 1290 Cl
Mulebrok 1220 LS, 1276 *Ass*, 1284 FA, 1287 *Ass*, 1302 FA
Millbrook 1227 Ass
Melebrok(e) 1247 *Ass*, 1253 Abbr, 1276, 1287 *Ass*, 1293 Ch,
     1316 FA
Mulbrok(e) 1291 NI, 1311 Ipm, 1364 IpmR, 1395 Cl
Melabroc 1291 Tax
Melbruk 1323 Cl
Melbroke 1330 Ipm
Milbrok 1346 FA, 1350 Ipm
Milebrok 1362 Cl
Milbrouke 1363 IpmR
Mullebr(o)ke 1366, 1395 Cl
Milbrooke 1428 FA
Temple Millbrook 15th HMC iv

Self-explanatory. Cf. *mylenbroc* BCS 675. The Knights Templars held a manor in Millbrook (1286 QW).

HAZEL WOOD (lost)

Hesilwode 1330 Ipm
Hazel Wood Lane 1766 J

Self-explanatory.

MOOR PLANTATION

la More 1330 Ipm

Self-explanatory.

Ridgmont

RIDGMONT [rigmənt] 95 A 5

Rugemund 1227 Ass
Rogemund 1276, 1287 *Ass*
Rychemund 1276 *Ass* (p)
Rugemunt 1286 Dunst, 1355 Cl
Richemond 1287 *Ass*
Rugemond 1316 FA
Rougemont 1349, 1356, 1368 Cl

de Rubeo Monte 1349, 1358 Cl
Rouge Mount 1368 Cl
Regmond, Regemont 1526 AD i, 1530 LP
Richmount 1527 LP
Ridgemond, Rydgemonde 1535 VE, 1549 Pat

A Norman-French name descriptive of the sandstone ridge on which the village stands. There is a definitely reddish tinge in many layers of this sandstone. Similarly, Rougemont, the hill of the Norman castle at Exeter, is named from the rich hue of the New Red Sandstone rocks (MS note of Professor Earle), while Mountsorrel (Lei) is named, as Mr Bruce Dickins reminds us, from the pinkish granite quarried there. The later phonological development into *Ridg*mont has probably been affected by a desire to associate the name with the very definite *ridge* which marks the site.

BECKERINGS PARK

Bickrings Park 1766 J

John de Bekeryng, who presumably came from Beckering (L), held one-fifth knight's fee in Segenhoe in 1346 (FA).

BROGBOROUGH

Brockeberg(h) 1222 FF, 1240 Ass, 1286 Dunst
Brokebergh, Brokesbery 1247 *Ass*
Brokeberwe 1261 Abbr
Brockeborewe 1308 Ipm
Brock(e)burwe 1324 Ipm, Cl
Brokkebergh 1328 Cl
Brok(e)burgh 1331 QW, 1388 Cl
Brockeburgh 1331 QW, 1354 Ipm
Broybury, Broybiry 1363, 1383 Cl
Brokboroughe 1389 IpmR, 1396 Cl
Brockborough 1509 AD v
Brobury 1525 LP

OE *brocc-beorg* or *brocca-beorg*, 'badger-hill' or 'badgers' hill.' It is tempting to take the name as from OE *brocen beorg*, 'broken hill' (cf. Brokenborough (W)), which would aptly describe its outline, but the fairly frequent *ck* and *kk* in early forms seem to make this impossible. *v.* **brocc, beorg.**

NORTH WOOD (lost)

Norwde 1193 FF
Nortwde 1250 FF

Self-explanatory.

SEGENHOE (Manor) [segnou] olim [sedʒnou]

Segenehou 1086 DB
Segenho(u) 1220 Ass, 1224 FF *et passim*
Sekenho 1234 (13th) *Dunst*
Seggenhou 1247 *Ass*
Seginho 1286 Dunst
Segonho 1346 FA
Sedgynhoe 1526 AD i
Segnoo, Segnow 1526 LS, 1576 Saxton, c. 1750 Bowen
Segnew 1527 LP
Sedgenow 1535 VE

'Segga's hoh or spur of land.' For this pers. name, cf.
Seckloe (PN Bk 16).

WINTER WOOD

Wintreho 1193 FF
Winterho 13th *Dunst*
Wintroe Corner 16th VCH iii. 321

'Wintra's hoh.' The pers. name *Winter, Wintra* is well
established in OE and lies behind Winteringham and Winterton (L), Wintringham (Hu, Y).

Steppingley

STEPPINGLEY 95 A 6

Stepigelai 1086 DB
Stepingelea 1167 P
Stepingele(gh) 13th *Dunst*, 1202 Ass, 1220 LS, 1276 *Ass* (p),
1302 Cl
Steppingele(gh) 13th *Dunst*, 1202, 1227 Ass
Stepyngle, Stepingle 13th *Dunst*, 1202 Ass, 1227 FF, Ass,
1276 *Ass*, 1284 FA, 1287 *Ass*, 1316, 1346 FA, 1390–2 CS,
1433 AD vi
Stebbingele(gh) 1214 FF, 1227 Ass

Stebbingle 1240 Ass
Steppingle 1242 Fees 867, 1323 Cl
Stypingle 1405 Inq aqd

This must be from OE *Stēapinga-lēage*, v. leah. There is
evidence for an OE pers. name *Stēapa*, the name being found
in an abstract of a late 10th cent. document entered in the *Liber
Eliensis*, a source which supplies good forms of other rare
pers. names. This would explain Steeping (L). There is
however an alternative possibility. Steppingley is on a well-
marked hill. An OE *Stēapingas*, 'people on the slope,' is a
possibility. The OE adj. *stēap* may have early been used as a
substantive in p.n. (cf. Steep (Ha), which goes back at least to
the 13th cent.) and an *-ingas* formation even from the adj. direct
is not impossible.

FROGHALL

Froghole 1853 O

Wilshamstead

WILSHAMSTEAD [wilstəd] 84 H 8

Winessamestede 1086 DB
Wilsamstud a. 1181 (13th) *Dunst*
Wyleshamstede 1220 LS, 1287 *Ass*, 1291 Tax
Wilchamstede 1239 FF
Wilshamstede, Wyl- 1240 FF, 1276 *Ass et passim*
Wilhamstede, Wyl- 1242 Fees 868, 1247, 1276 *Ass*
Wilsamstede 1247 *Ass*
Wilhampstede 1372 Cl
Wyllsamsted 1526 LS
Wilsumstead 1675 Ogilby
Wilstead 1780 Jury

'Wil's homestead' v. hamstede. Cf. Zachrisson, *Some
English PN Etymologies* 4. The weak form *Willa* is well established. It looks as if there were also a strong form *Wil* formed
directly from the first element of such names as *Wilbeorht*.

The DB form stands alone, and as AN confusion of *l* and *n* is
fairly common it should probably be rejected.

I

WILSHAMSTEAD COTTON END (6″)

Westcote, Westcota 1086 DB *et passim*
Cotes 1228 FF, 1284 FA
Cot(e)nes 1247 *Ass*
Westcotes 1247, 1287 *Ass*
Welleshamsted cum Wescote 1287 *Ass*
Wescote juxta Wyleshamstede 1287 *Ass*

The various forms of the second element of this name can be explained as in EPN *s.v.* cot, except for *Cotenes* which is from the dat. pl. coten (OE cotum) with the plural suffix *es* added when the name was not understood. *West* not in relation to the parish but to Cardington Cotton End. *v. infra* 91.

Wootton

WOOTTON 84 G 6

Otone 1086 DB
Wutton 1197 FF *et passim* to 1397 Cl
Wotton 1223 FF *et passim*
Wuttun Hy 3 (1317) Ch
Woutton 1304 Ch

'Wood-farm' *v.* wudu, tun.

HALL END

Hallend Edw 3 (15th) *Newn* 142

Self-explanatory.

HOO (Fm)

le Hoo 1242 Fees 885
Hoo 1276 *Ass*
Ho 1303 *KF* (p)
Wottonhoo 1400 IpmR

v. hoh.

KEELEY LANE

Gyldelewode 1446 IpmR

The identification is not certain. If it is correct we may compare, for confusion of initial *c(k)* and *g*, the history of Catsey Wood *supra* 35.

WOOTTON BOURNE END

Wuttonebourne, Wuttebourne, hamelettum de 14th (15th) *Newn* 137

Bone End 1826 B

So named from the little stream (*v.* burna) which forms the boundary of Wotton parish on the north-west.

WOOTTON BROADMEAD

Brodemade 14th (15th) *Newn* 139

v. mæd. Self-explanatory.

WOOTTON PILLINGE [pilidʒ]

Driepelenge 1284 *Newn* 53
Peling(e), Pelyng, Pellyng, Pellinge 1287 *Ass* (p)
Peling in Wottone 1311 Ipm, 1347 BM
Pelynge 1434 IpmR
Pelyng 1445 IpmR
Wotton Pillage Eliz ChancP

One cannot carry the name of this place (and that of the neighbouring Pillinge in Marston) further than is suggested by Ekwall who (PN in *-ing* 162) derives it from an OE pers. name *Peol, Piol,* evidenced in *Pioles clifan* (BCS 1282), Pelham (Herts) and Pelsall (St) which is found in the Wolverhampton Charter, Dugd. vi. 1444, as *Peoleshale* together with a *Peolesford.* If this is its history it is to be taken as a formation in *-ingas* (cf. EPN 41 foot) and means '(settlement of) Peola's people.' For the pronunciation, cf. Lymage *infra* 270. In the first reference the place is called 'dry.'

V. WIXAMTREE HUNDRED

Wichestanestou, Wichenestanestou 1086 DB
Wixtanestre 1162 P
Wic(h)stanestre 1163, 1169 P, *Wich Stanestre* 1175 P
Wicstanistre 1173, 1175 P
Wichstonestre 1176 P
Wixtonestre, Wyxstonestre 1183 P, 1287 *Ass,* 1316, 1346 FA
Wikestonestou 1185 P
Wicstanestan 1191 P
Wistanestre 1202 Ass, 1276 *Ass*

Wykestanestre 1247 *Ass*
Wyston(e)stre 1284 FA
Wyxtontre 1428 FA
Wixantre 1549 Pat

OE *Wihstānes-trēo*, 'Wihstan's tree,' the reference being to
the tree by which the Hundred met. This name is found in the
forms *Wihstan, Weohstan, Weoxtan* in ASC. For such tree-
names, *v.* Mawer, *PN and History* 23, and cf. Willey Hundred
supra 25. For the substitution of stow for treo cf. Redbornstoke
Hundred *supra* 66 and cf. Bunsty Hundred (Bk) which has the
same suffix. The site of the Hundred meeting-place is unknown.

Blunham

BLUNHAM 84 F 11

Blunham 1086 DB, 1220 LS, 1227 Ass, 1247, 1276, 1287 *Ass*,
　　1309 Ch *et passim*
Bluneham 1086 DB, 1240 FF, 1247, 1303 Cl
Bluham 1227 Ass
Blowenham 1250 Pat, 1287 *Ass* (p)
Bluenham 1276 *Ass*
Blounham 1276 *Ass*, 1287 *Ass* (p), 1316 FA, 1325 Ipm, 1331
　　Fine, 1345 Cl, 1346 FA, 1347 Cl, 1355, 1360 Ipm, 1376,
　　1395 Cl, 1401 IpmR, 1428 FA, 1517 AD vi, c. 1550 *Linc*
Blonham 1276 *Ass*, 1584 AD v
Blownham 1314 Ch, 1526 LS
Blownam 1425 BM
Bleweham 1527 BM
Bloneham 1531 BM, c. 1550 *Linc*
Bloungham 1629 BM

v. ham. The first element is probably a pers. name. Professor
Ekwall agrees that no certainty can be attained but suggests the
possibility of an OE *Bluwa* cognate with the pers. name found
in OHG *Pluwenhofen, Pluvileshusirum*.

Cardington

CARDINGTON [kæriŋtən] 84 F/G 9

Chernetone 1086 DB
Kerdinton, Kerdynton c. 1190 HMC, Var vii, 1227 Ass, 1247
　　Ass, 1254, 1257 FF, 1274, 1302, 1338 Ipm, 1428 FA

Kardinton 1220 LS, 1227 Ass
Kerdintuna Hy 3 (1304) Ch
Cherdyntona Hy 3 (1317) Ch
Kerdington, Kerdyngton 1247 *Ass*, 1276 RH, *Ass*, 1287 *Ass*,
   1304 Ch *et passim* to 1428 FA
Kerenton al. *Kerynton* 1286 Ipm
Kardington, Cardyngton 1287 *Ass*, 1440 IpmR
Carington 1535 VE, 1549 Pat
Careington 1638 Cai

'Cærda's farm' *v.* ingtun. This name, like Cardington (Sa),
Cardwell(D), seems to contain the same pers. name that is found
in Charndon and Chartridge (Bk), but the absence of palatalisation is difficult. For this pers. name cf. *Cardanhlæw* (BCS 877,
orig. 10th cent. charter), *Cærdan-, Cerdan-hlæw* in Berks (BCS
963, 12th cent. MS). For the possibility of forms with and
without palatalisation *v.* the evidence for *Card-, Cerd-* names set
forth by Stevenson in EHR xiv. 34 n. Professors Ekwall and
Zachrisson would prefer to derive this from OE *Cēnrēdingtūn*,
through an assimilated form *Cerredingtun*. In this case the early
Card- forms offer difficulties.

CARDINGTON COTTON END

  v. Eastcotts *infra* 90.

Cople

COPLE 84 F 9
  C(h)ochepol 1086 DB
  Cogopol c. 1150 BM
  Coggepol(e) c. 1150 (c. 1230) *Warden* 84, 1196, 1202, 1207,
    1211 FF, 1227 Ass
  Coggispol 1195 P
  Cogepol 1202 FF, Hy 3 BM
  Cokepol 1227 Ass
  Cokkepol, Cockepol c. 1230 WellsL
  Cowepol 13th (15th) *Newn* 186
  Coupel 1240 Ass, 1247, 1276 *Ass*
  Coupol(e) 1247 *Ass*, FF, 1276, 1287 *Ass*, 1291 NI, 1307 *Ass*,
    1311 Ipm, 1316 FA, 1327 Cl, 1347 BM, 1382 Cl
  Caupel, Caupol 1247, 1276 *Ass*

Coupul 1361 Cl
Coupulle 1388 Cl, 1428 FA, 1434 IpmR
Coupell 1402 BHRS i. 103
Coupehill 1461 IpmR
Copull or *Coupull* 1509 LP
Coople c. 1530 *Linc*
Cowple 1535 VE
Cowpull 1535 VE
Copull 1549 Pat

It would seem possible to explain this difficult name if we assume that the man from whom the pool took its name (*v.* pol) was called *Cogga*, a name found in *Cocgganhyll* (So) BCS 670, *Cogganbeam* (Ha) BCS 1200, and *Cogan mære* (Ha) BCS 1080, and (in a strong form) in Coggeshall (Ess) and also Cogswell *infra* 98. The DB form and the isolated form with *k* in 1227 could be explained naturally by unvoicing of *g* to *k* before *p*. This pers. name *Cog(g)a* is presumably related to OE *Cugga*, which may be inferred from Cogenhoe (Nth), *Cughanhocg* (12th cent.).

Alternatively, but in view of the forms less probably, we might start from a personal-name *Cocca* such as is found in Cockbury (Gl), *Coccanburh* BCS 246, and then take the *g(g)* forms to show the same voicing of intervocalic *k* which is illustrated in Eggington *infra* 121 and possibly in the adjacent Moggerhanger. Cf. also Cogshall (Ch), DB *Cocheshull*, 1287 *Cogeshull*. Professor Ekwall suggests further that just as stopped *d* may become continuant *ð*, and *b* become *v* (cf. Pavenham *supra* 36), so stopped *g* might become a continuant *g* and then quite regularly form a diphthong with the preceding *o*. This would explain the later *ou* forms. For the combination of *pol* with a pers. name cf. Widmerpool (Nt), where *pol* is preceded by OE *Widmǣr*.

COPLEWOOD END

boscus de Coupel 1276 *Ass*

Eastcotts

EASTCOTTS [i·stkəts] (6″)

Cotes 1220 LS, 1227 Ass
Estcotes 1240 Ass, 1393 Cl

Escotes 1382 Cl
Escotts 1780 Jury

'The parish takes its name from the "cotes," now represented by Cotton End in Cardington, called "east" in distinction from the "cotes" in Wilshamstead, once known as *Westcotes* (*v. supra* 86) now as "Wilshamstead Cotton End."'

In this parish we have mention in the Newnham Cartulary (189) of a *Cotingebroc*. This is an interesting illustration of the use of the suffix ingas to denote the inhabitants of a place, for it is clear that this is from OE *Cotingabrōc*, 'brook of the dwellers at the cotes.'

EXETER WOOD

This owes its name to the fact that one of the Cardington manors was in the hands of Lord Exeter's family from 1577–1879 (VCH iii. 235).

FENLAKE

Fenlak(e) 1247, 1276 *Ass*

'Marsh stream' *v.* fen, lacu.

HARROWDEN

Herghetone, Hergentone 1086 DB
Harewedon 1166 P, 1202, 1227 Ass, 1276 *Ass*
Harwedon 1227 Ass, 1247 *Ass*, 1393 Cl
Harwdon 1247 *Ass*
Harudon 1276 *Ass*
Haroudon 1276 *Ass*
Harouden 1287 *Ass*

This would seem to have the same history as Harrowden (Nth) and to be a compound of hearg and dun. If so it must have been, like Harrow (Mx), a place for heathen worship in ancient days, with a sacred grove or shrine. Cf. Harrowick *supra* 14.

Moggerhanger

MOGGERHANGER [mɔˑrhæŋə] 84 F 10

Mogarhangre 1216 Cl
Moggehangre 1220 LS

Mogerhanger 1240 FF, 1247 *Ass*, 1262, 1270 FF, 1276 *Ass* (p),
    1284 FA, 1289 Ipm[1], 13th AD iii, 1316 FA, 1327 Ipm, 1346
    FA, 1351 BM, 1358 Cl, 1428 FA
Mokehangre 1242 Fees 870
Moukeraungre 1276 *Ass* (p)
Mokerhanger, *-re* 1276, 1287 *Ass*, 1290 Cl, 1370 IpmR,
    1389 Cl, 1398 IpmR
Mougerhanger 1287 *Ass*
Mogerhanger al. *Mouerhanger* 1289 Ipm
Moggurhangger 1347 AD vi
Mokerangre 1394 Cl
Moker Aungre 1394 IpmR
Mogranger 1474 IpmR, 1535 VE
Mogreangre 1488 Ipm
Mogeranger 1517 AD vi, 1535 VE
Mowgranger 1629 BM
Morehanger 1675 Ogilby
Muggeranger c. 1750 Bowen
Muggerhanger 1780–1830 Jury *passim*

The second element in this name is hangra. The first offers
great difficulties. Pursuing the same line of phonological development which was suggested under Cople, Professor Ekwall
would start from a name beginning with *Moker-* or *Muker-*,
take the *g*(*g*) forms as due to intervocalic voicing, and the
Mouer- of 1289 and the modern pronunciation as *Morehanger*
to be due to continuant *g* developing from earlier stopped *g*.
For *Muker-* itself he suggests possible affinities with ME *muke-
ren*, 'to hoard,' whence perhaps *muker*, 'miser,' though the
regular form seems to be *mukerer*. Was it possibly 'misers'
wood' from an ancient hoard discovered there?

The local pronunciation has been preserved in Morhanger
House in the parish of Moggerhanger (cf. the reverse process in
Attingham Hall in Atcham (Sa), where the name of the Hall
preserves the fuller form).

CHALTON

Cerlentone 1086 DB
Cherleton 1173 P, 1242 Fees 870, 1331 QW, 1428 FA

---

[1] Writ endorsed *Morhanger* in a contemporary hand (G.H.F.).

Cherlton 1227 Ass (p)
Charleton 1240 FF, 1287 *Ass*, 1535 VE
Chelton 1242 Fees 894
Chauton 1250 Ch
Chaltone 1276 *Ass*, 1333 Ipm, 1369 Cl
Chaulton 1766 J

v. ceorl, tun.

SOUTH MILLS

Sudmille 1220 LS (p)
South Mulne 1245 Ipm
Suth Mulne 1270 FF
Suthmulle 1276, 1287 *Ass*
Suthmilne 1276 *Ass*
South Milne 1361 Cl

Self-explanatory.

Northill

NORTHILL [nɔrəl]
Nortgiuele, Nortgible 1086 DB
Northgille 1185 P
Norttgifelle 13th *Dunst*
Northgiuel(e), -gyu- 1219 FF, 1227, 1240 Ass, 1261 FF, 1294
  Ipm
Nortgylle 1242 Fees 885
Nortgiuel 1247 *Ass*
Northgevle 1271 Ch
Northyeuil, Northiuel 1287 *Ass*
Northyevel(e) 1290 Ipm, 1301, 1313 Cl, 1314 Fine, 1316 FA,
  1317 Ipm, 1319 Cl, 1346 FA, 1379 AD vi, Cl, 1387 Cl
Nortgewele, Norgivill 13th AD vi
Northyvele 1303, 1364 AD vi
Northevele 1346 FA
Noryevele 1346 AD vi
Norrell al. *Northyevell* n.d. AD vi
Norrell 1443 AD vi, Eliz ChancP
Northyevyll 1470 AD vi
Northyell 1494 BHRS ii. 123
Northewell 1526 LS

Norryell, Northyell 1535 VE
Northiell 1536 BM
Norrihill 1537 AD v
Northwell al. *Norrell* al. *Northeywell* 1549 Pat
Norhill 1675 Ogilby

Northill and Southill *infra* 96 alike contain as their second element the name of the river Ivel, as shown by the forms of that river-name given above, though each village is a good two miles to the west of the river. That river-name is however certainly the source of the territorial name *Gifla* noted above (*v.* Introduction xviii), and Northill and Southill may be descriptive of settlements in the north and south of that territory rather than loose appellations for villages which are well away from the river itself. For the modern pronunciation we may compare Norham (Nb).

BROOK END

le Broc 13th AD iii
le Brokende 13th AD vi

Self-explanatory.

BUDNA

Budenho(u) 1169 P (p), 1202, 1227 Ass, 1232 FF *et passim*
Buddenho 1195 P (p), 1276, 1287 *Ass* (p)
Bodenho 1276 *Ass* (p), 1390–2 CS (p)
Boddenho 1297 *SR* (p), 1344 AD vi
Buddinho 1337 AD vi
Bodynho 1416, 1424 AD vi
Budno E 4 AD iii
Budenow al. *Bodenho* 1519 AD vi
Budnow 1539 AD v
Budnall, Budnoo 1549 Pat

'Bud(d)a's hoh' or spur of land, *Bud(d)a* being a well-established OE name.

CALDECOTE (Upper and Lower) [ka·kət]

Caldecot(e) 1197 FF, 1227 Ass *et passim*
Kaldekote c. 1200 (c. 1230) *Warden* 81 *b*
Caudecot(e) 1202 FF, 1276 *Ass* (p)
Caldecote, Magna et Parva 1234 Cl

Nethere Caldecote 1351 AD iii
Overe Caldecote 1363 AD ii
Calcut 1513 AD vi

'Cold cottages,' so called from their exposed situation. *v.* ceald, cot.

COLLEGE POND and WOOD

This is the last trace of the fact that Northill Parish Church became a collegiate church in 1404 (Pat).

HATCH

la Hache 1232 FF, 1247 Ass
la Hacche 1247 *Ass*
Hattche 1539 AD vi
cf. *le Hacchedich* (1259 AD vi)

v. hæcc. The compound *hacchedich* found in the same area suggests that the hatch here was some kind of floodgate or sluice.

HILL LANE

le Hul 1460 AD ii

Self-explanatory.

ICKWELL

Ikewelle c. 1170 (c. 1230) *Warden* 82
Gikewelle, Gyk- c. 1180 (c. 1230) *Warden* 82, 1202 Ass, FF,
    1247 *Ass*, 1261 FF, 1276 RH, *Ass*, 1286 Dunst
Chikewelle, Chyk- 13th *Dunst*, 1202 Ass, 1260, 1262 FF
Gigewell 1202 Ass
Gikeswell 1227 Ass
Jekewelle 1240 Ass, 1346, 1428 FA, 1457–8 BM
Yikewell 1287 *Ass*
Yekewell 1287 *Ass*, 1434 IpmR, 1457–8 BM
Gekyewelle 1360 AD iii
Zykwell 1379 AD vi
Zekewell 1400 IpmR
Ikwell 1552 AD vi

The early forms of this name point clearly to a pers. name *Gica* or *Gicca*. For the possibility of such a name reference may be made to the history of Kickle's Farm (PN Bk 22). Hence 'Gic(c)a's stream or spring' *v.* wielle.

THORNCOTE

Thornecote 1206 FF *et passim*
Thurnecot n.d. AD vi
Thurnekote 1300 AD iii
Thorncut 1780 Jury

'Cote(s) by the thorn-bush' *v.* þorn, cot.

Shefford Hardwick

SHEFFORD HARDWICK 84 J 18

Herd(e)wik 1227 Ass
Herdewich 1227 Ass
Herdwyk juxta Sheford E 2 Orig

v. heordewic. It is clearly a pastoral settlement from the manor of Shefford.

COLLINS GROVE

Collins Grove 1374 AD i

Southill

SOUTHILL [sʌðil] 84 H 11

Sudgiuele, Sudgible 1086 DB
Sudgill 1197 FF
Sutgivel(e) R i (1286) Ch, 1214 FF
Sugivel(e), Suȝiuele, Sugyvel 1215, 1219 FF, Hy 3 BM
Sutgylle 1242 Fees 885
Suthgiuel, Suthgyvel c. 1200 (c. 1230) *Warden* 87 b, 1219, 1247 FF, 1282, 1329 Cl
Suthyeu(e)le 1227 Ass, 1276 *Ass*, 1316 FA, 1327 Cl
Suggivel 1229 Cl, 1247 *Ass*
Suggeuel 1247 *Ass*
Sut(h)yvell 1273 Ipm, 1287 *Ass*
Sugul, Sugil 1276 *Ass*
Southyevel(l) 1338, 1342, AD v, 1346 FA, 1381, 1388 Cl, 1391 AD v, 1399 Ch, 1428 FA, 1451 BM
Southyevill 1400 CS
Southiell, Southyell 1515 LP, 1535 VE, 1548 Pat
Southwell 1518 *Award*, 1582 Cai

For the explanation of this name, *v.* Northill *supra* 93.

BROOM

Brume 1086 DB

Brome Hy 2 (c. 1230) *Warden* 94, 1202 Ass

So called from the abundance of the plant. *v.* **brom.**

FOXHOLE COVERT

Overfoxehole 13th AD ii

Self-explanatory.

GASTLINGS

Katelynesbury 1367 Orig

Gatelynesbury 1370 IpmR

Gaslins 1648 NQ iii

This is a manorial name (*v.* burh). Already in 1250, Sir Geoffrey Gacelin held land in Southill (Ipm). The *t* in the first two forms may well be a mistake for *c*.

ROWNEY

Rueye R i (1286) Ch

Runheye 1252 Ch

Roun(h)ey 1291 Tax, 1297 *SR* (p)

Roundhey 1535 VE

This is from OE *rūhan* (*ge*)*hæge* (dat.), 'rough enclosure,' *v.* **gehæg.** This suits the locality, which is a wooded one. It is definitely not an island site (**eg**) even in the widest sense of that term.

STANFORD

Stanford 1086 DB *et passim*

Stamford 1202 Ass, 1247 *Ass*, 1377 Cl

Staunford 1227 Ass, 1238 FF, 1247 *Ass*, 1346 FA

Standford 1535 VE

Standfordbury 1588 D

'Stony ford.'

Old Warden

OLD WARDEN 84 H 10

Wardone 1086 DB *et passim*, c. 1350 *Linc* (*juxta Bedeford*)

Wardun 1158 P, c. 1170 (c. 1230) *Warden* 83, 1215 BM, 1239 Orig, 1244 Ch, 1444 Orig

Waredon 1227 Ass, 1229 Cl, 1262 FF
Warden 1359 Ipm
Wardoun 1383, 1388, 1393 Cl
Old Wardon 1495 Ipm
Worden 1785, 1798 Jury

Skeat must be right in his suggestion that this is for OE **weard-dun**, 'watch-hill.' Cf. Warden (Nb) for a similar name.

BROOKLAND (Fm)

cf. 1422 AD vi *le Brooke*

Self-explanatory.

COGSWELL (lost)

Kokeswell 1291 Tax
Kakeswell 1407 AD vi
Coggeswell 1535 VE

It is difficult not to believe that this place was named from the same person whose name is preserved in a weak form in the adjacent village of Cople *supra* 89. It tends to support the view that *Cocca* rather than *Cogga* is the original form of that name.

HILL (Ho)

Hulle 1276 *Ass* (p)
Hylle 1422 AD vi, 1433 AD iii
le Hill 1440 AD v

PARK (Fm, Wood)

Grangia de Parco R i (1286) Ch

WARDEN STREET

Wardon in le Strete 1549 Pat

WARDEN (GT and LITTLE) WOOD

boscus de Wardon 1287 *Ass*

The last four names are self-explanatory, except for Warden Street. The *strete* on which this hamlet of Warden lies is probably that present combination of foot-path, cart-track and road which makes its way straight to Moggerhanger and thence by

Blunham to the Ivel, crossing it near Tempsford. Reference to the southward continuation of this road seems to be made in the names *Heyestrate, Stretfurlong* in Southill (*Newn* 199 *b*). This is one of those names in which *street* is used of a road apparently not of Roman origin.

Willington

WILLINGTON 84 F 9

Welitone 1086 DB

Willitona, Wyll- c. 1150 BM, 1276 *Ass*, 1376 IpmR

Wilitone, Wyl- 1202 Ass (p), 1227 Ass, 1247, 1287 *Ass*, 1299 Cl, 1307 *Ass*, 1323 AD i, 1326 Ch, 1382 Cl, 1385 Ch, 1388 Cl

Wylinton, Wil- 1220 LS, 1227 Ass, 1276 *Ass*

Wulinton 1227 Ass

Wiletuna Hy 3 (1317) Ch

Wylyton 1276 *Ass*

Wel(l)inton 1276 *Ass*

Wylington 1276 *Ass*, 1327 Ipm, 1328, 1361 Cl

Wilton 1284 FA, 1287 *Ass*, 1316 FA, 1322 AD i

Welyngton 1354 Cl, 1428 FA

Wyllington, Will- 1457 BM, 1488 Ipm, 1539 AD v

This at first sight suggests the same personal name *Wila* which we have already found in Wilden *supra* 66, hence 'Wila's farm' (*v.* ingtun), but the earliest forms point rather to OE *welig-tun*, 'willow-farm' (*v.* welig), with later intrusive *n*.

MILL FARM

molendinum de Wiliton 1359 (15th) *Newn* 20 *b*

SHEERHATCH WOOD

Shirhacche 1369 Orig

Shirehache 1407 IpmR

Shire Hatch Wood 1826 B

The 'hatch' (*v.* hæcc) was probably a gate into the wood. The first element is presumably the OE pers. name *Scīra*, though it is just possible that it may be the adj. scir used to describe a brightly-coloured gate. The modern form is clearly corrupt.

VI. BIGGLESWADE HUNDRED

Bichelesuuorde, Bicheleswade 1086 DB
Bicheleswatere 1107 (c. 1200) Hist Mon de Abingdon ii. 101

For other forms, *v*. Biggleswade *infra* 101. Biggleswade itself lies centrally for the half-circle of parishes which form the Hundred.

Half-Hundred

WENSLOW

Weneslai 1086 DB
Wodneslawe 1169 P
Wodnislawe 1175, 1179 P
Wendelawe 1202 Ass
Wadeslowe 1247 *Ass*
Weneslawe 1287 *Ass*

This half-hundred clearly took its name from a hill or barrow (*v*. hlaw) sacred to the worship of Woden. Similar hills or barrows (OE beorg) sacred to Woden are Wednesbury (St), locally [wedӡbəri], and Woodnesborough (K), locally [winzbəri]. The half-hundred included Everton, Potton, Hatley, Sutton and Sandy and is now absorbed in Biggleswade.

Astwick

ASTWICK [ɑ·stwik] 84 J 13

Estuuiche 1086 DB, 1175 P (p)

Forms with *Est-* alone are found until

Hastewik 1247 *Ass*
Astwyk 1316 FA, 1349 BM, 1369 Fine, 1389 AD ii

but *Est-* forms are occasionally found until

Estwyk 1436, 1443 BM

'The eastern **wic** or farm,' so called because it lies on the eastern border of the county.

Little Barford

LITTLE BARFORD 84 D 12

Bereforde 1086 DB, 1167 P
Berkeford 1202 Ass, 1243 FF, 1247, 1287 *Ass*, 1318 Ipm, 1409 AD i, 1480 IpmR, 1581 BM

Berck(e)ford 1220 LS, 1291 NI

Bercford 1247 *Ass*, 1250 FF

Berkford 1269 FF, 1287 *Ass*, 1297 *SR*, 1316, 1346 FA, 1355 BM, 1359 AD iv, 1387, 1393 Cl, 1390–2 CS, 1428 FA, 1576 Saxton

Berford 1284 FA

Berecforde 1346 Rams

Barkford 1415, 1509 BM, 1526 LS, 1554 FF, 1586 D, 1748 BHRS ii. 147

Barford 1539 BM

Barkford al. *Barford* 1549 Pat

OE *beorca-ford*, 'ford of the birch-trees,' *v.* beorc, ford. As the forms show, the original forms of Barford in the two Beds parishes of that name were quite different and we are not surprised therefore to find that the distinction of them as *Great* and *Little* Barford is of modern origin.

Biggleswade

BIGGLESWADE 84 G/H 12

Pichelesuuade 1086 DB

Bicheleswada 1132 *D and C Linc* A 1/1 no. 2, 1202 Ass

Bic(h)leswade 1175 P, 1227 Ass, 1227–9 Ch, 1229 Cl

Bikeleswade 1183 P, 1202 Ass, 1220 LS

Bickeleswade 1240 Ass, 1255 *Bodl* (*Bk* 2)

The forms run on these lines except for

Becleswade 1438 AD iv

Beklyswade 1480 BM

until

Bygelswade 1486 AD ii

Bygleswade 1535 VE

but forms with *c* or *k* are still common as in

Biccleswade al. *Biggleswade* Eliz ChancP

'Biccel's ford' *v.* wæd. This pers. name is not found on independent record in OE but is found in Bigstrup (Bk). It is a diminutive of OE *Bicca* and has a cognate in OGer *Bichilo*.

K

HOLME

Holme 1086 DB *et passim*
Hulmus, Hulme 1179 P, 1198, 1199 Cur, 1211 FF, 1260 FF
(p), 1276 *Ass*
Ulmus 1199 FF
La Hume 1219 FF

v. holmr. It is almost surrounded by streams. The frequency of forms with *u* rather than with *o* is interesting as it points definitely to Danish rather than Norse influence. Cf. IPN 60.

SCROUP'S FARM (6")

Henry le Scrope died seized of land in Holme in Biggleswade in 1337 (Ipm).

SHORTMEAD (Ho)

S(h)ortemade 1276, 1287 *Ass*
Schortmede 13th AD ii
Schort(e)made 1311 AD vi, 1333 Ipm

The name is self-explanatory and is very common as a field-name.

STRATTON

Stratone 1086 DB
Stretune 1107 (c. 1200) Hist Mon de Abingdon ii. 100
Strattune 12th *HarlCh* 83 B 39 (p), 1199 FF
Straittune 12th *HarlCh* 83 A 47
Stratton 1202 FF *et passim, under Hungurhul* 1337 AD i, *by Bykeleswade* 1383 Cl
Stretton 1247 FF, 1331 QW, 1339 BM, 1393 Cl, 1766 J

'The farm on the stræt.' This term is applied to the Roman road, also known as the White Way, which comes up from Baldock in Herts and makes its way ultimately to Godmanchester (cf. VCH ii. 4). The 'hunger' hill is presumably a term of reproach for the barren hill which slopes up from Stratton Farm (150 ft.) to a point on the Roman Way (225 ft.) which, significantly enough, is near a 'Bleak' Hall.

Dunton

DUNTON 84 H 13

Donitone, Danitone 1086 DB
Duniton 1185 Rot Dom
Dunton E 1 (1286) Ch *et passim*, (*by Bikeleswade*) 1333 Ipm
Donton 1202 Ass, 1328 Ipm, Fine, 1504 Ipm
Dutton 1242 Fees 869
Dounton 1287 *Ass*, 1316 FA, 1328 Ipm (*Chaumberleyn*), 1400
   CS, 1404 IpmR
Danton Eliz ChancP

The earliest forms with medial *i* are a little disconcerting, but in view of the uniform character of the later ones and the clear topography of Dunton, lying on a ridge, we must interpret the name as from OE *dun* and *tun* and take it to mean 'hill-farm.' The Chamberlain family held land here as early as 1210 (VCH ii. 212) and were the same family that gave their name to Compton Chamberlayne (W). *v.* Addenda.

MIDDLESEX (Fm)

Myddelsex 1444 AD ii

The solution of this difficult name is perhaps to be found in an entry in *Miscellaneous Inquisitions* (1327)[1]. In that year Robert of Baldok, archdeacon of Middlesex, died seized of the manor of Stanbridge in virtue of a fine. The manor ought to remain to Richard son of John le Chamberleyn and Margaret his wife. Stanbridge is on the other side of the county, but as the manor of Dunton was in the 14th cent. also in the possession of the Chamberlain family it seems possible that this manor also may have come into the possession of the archdeacon of Middlesex and that he left his name in this farm.

MILLOW

Melnho 1062 (12th) KCD 813, 1202 Ass, FF, 1247, 1276,
   1287 *Ass*, 1307 Ipm, 1331 QW, 1363 Cl, 1401 IpmR
Melehou 1086 DB
Mul(e)nho 12th c. HarlCh 83 B 39 (p), 1276, 1287 *Ass*, 1316
   FA, 1342 AD vi, 1346 FA, 1351 Ipm, 1428 FA
Melho 12th c. HarlCh 83 A 47 (p), 1253 Ch

---

[1] For the solution we are indebted to Miss E. G. Withycombe.

Milcho 1201 Cur
Mulho 1204 FF, 1227 Ass
Miln(e)ho 1247, 1276, 1287 *Ass*, 1307, 1372, 1394 Cl, 1423
    BM, AD i, 1552 BM, Eliz ChancP
Milho 1276 *Ass*
'Mill-hill' *v.* hoh, myln.

NEWTON

Newtonbury 1504 Ipm
'The manor house by the new farm,' the manor being that
otherwise known as *Chamberlainesbury* from its holders.

Edworth

EDWORTH 84 J 13
Edeuuorde 1086 DB
Eddewrþe, Eddewrth 1198 FF (p), 1232, 1247 FF
Eddewurþe 12th *HarlCh* 83 B 39 (p)
Eddeworth 1202 Ass, E 1 BM, 1276 *Ass*, 1295 Cl, 1297 Ipm,
    SR, 1299 Ipm, 1306 Ch, 1307, 1315 Ipm, 1333 AD iii,
    1346, 1428 FA
Edewrth 1202 Ass
Eddewurth 1227 Ass, 1242 Fees 882
Edeworthe 1276 *Ass* (p), 1324 Ipm, 1325 Cl
Edisworth 1276 *Ass* (p)
Edesworth 1284 FA
Edeneworth 1315 Ipm
Edworth 1355 Cl *et passim*
'Edda's enclosure' *v.* worð. This is a pet-form for an OE
name in *Ēad-*, as illustrated in the passage quoted by Redin (65)
from Simeon of Durham, where we have '*Edwine*, qui et *Eda*
dictus est.'

Everton

EVERTON 84 E/F 12
Euretone 1086 DB, 1220 LS
Euretune 1086 DB
Euerdon 1227 Ass, 1276 *Ass*, 1355 BM
Euerton 1227 Ass *et passim*
'Boar-farm' *v.* eofor, tun.

Eyworth

EYWORTH 84 G 14

   Ai(ss)euuorde 1086 DB
   Eiwrth, Eywrth 1202 Cur, 1227 Ass, 1276 *Ass*, 1290 Cl
   Eyworth 1232 Cl, 1247 *Ass et passim*
   Eywurth 1235 Cl, 1247 *Ass*, 1254, 1256 FF
   Eyworth or *Eywood* 1343 Ipm
   Ayworth 1518 BM

   'Island enclosure,' as explained by Skeat. *v.* eg, worð. The
village has water on three sides. The DB form *Aisseuuorde* is
corrupt.

THISTLEYGROUNDS (Fm)

   cf. *Blakethistel* 1256 FF

   The field-name and farm-name are probably to be associated
and are self-explanatory.

Cockayne Hatley

COCKAYNE HATLEY [kɔkin ætli]

   Hattenleia c. 960 (13th) BCS 1062
   Hættanlea c. 960 BCS 1306
   Hætlea c. 1053 KCD 920
   Hatelai 1086 DB
   Hatteleg(h) 1172 P, 1227, 1240 Ass
   Hatele(ia) 1198 FF, 1247 *Ass*
   Hattele 1227 Ass, 1232 FF, 1240 Ass, 1297 *SR et passim* to
      1450
   Burihattele 1276 *Ass*
   Beriattele 1360, 1394 Cl
   Hatlee 1372 Cl
   Hatley 1487 Ipm
   Buryhattley 1499 Ipm
   Hatleyport 1535 VE
   Cocking Hatley 1576 Saxton
   Hatley Port or *Cockayne* 1671 BM

   'Hætta's clearing' *v.* leah. The OE form seems to make the
existence of this pers. name certain, but it is not otherwise
on record. Presumably it is a pet-form for one of the com-

pound names in *Hǣð-*. The Port family were connected with Hatley as early as 1197. It passed to the Cockayne family in 1417 (cf. VCH ii. 215). 'Bury' is used in its manorial sense (*v.* burh) but is prefixed rather than, as usual, suffixed. For *Cocking*, cf. Wilden *supra* 66.

Langford

LANGFORD 84 H/J 12

Longaford 944–6 (c. 1250) BCS 812
Langeford 1086 DB, 1220 LS *et passim* to 1399 Cl
Langford 1428 FA
Langforth 1498 Ipm

Self-explanatory.

VINE FARM (6″)

Vine Farm occupies an isolated position at the extreme corner of Langford abutting on Astwick, and the VCH (ii. 271) points out that this is the last trace of a vinery belonging to the manor of Astwick, mentioned in a 15th cent. lease.

Potton

POTTON 84 F 13

Pottun c. 960 (13th) BCS 1062, c. 960 BCS 1306, n.d. KCD
    1352, 12th cent. *D and C Linc D* ii 90/3 no. 22, 1241 Cl
Potone 1086 DB, Hy 2 (1329) Ch
Potton 1203 FF *et passim*
Potton juxta Sutton 1384 IpmR

This would seem clearly to be 'pot-farm,' but in what sense we cannot be sure. It may be so because pots were once made there. *pot* is also used topographically to denote a deep hole or pit. Such an application is unsuitable here (unless one can apply it to a very wide and shallow depression) and the word in this sense seems to be confined to the North Country. The only use of the word in OE charters is in *pottaford* (Sf) BCS 1269. This is ambiguous as it may mean 'ford of the (broken) pots,' just as much as 'ford with the holes.' Potcote (Nth), if we may judge by the early forms, similarly, does not contain a pers. name.

Sandy

SANDY 84 F 11

Sandeie 1086 DB, 1185 P *et passim*
Sandun, Sandon 1197 FF, 1206 FF (p)
Sandee 1202 Ass
Saundee 1227 Ass
Saundeye 1227, 1240 Ass, 1247 *Ass*, 1251 FF, 1276 *Ass*, 1294
  Ipm, 1297 *SR*, 1309 Ch, 1325 Ipm, 1338, 1358, 1368 Cl,
  1376, 1399 IpmR
Sondy 1247 *Ass*
Sandheye 1287 *Ass*
Sondey(e) 1287 *Ass*, 13th AD vi, 1316 FA, 1372 BM, 1394,
  1395 Cl, 1400 CS, 1422 IpmR, 1428 AD vi, 1489 Ipm
Sonday 1507 BM, 1547 Pat
Sondheye 1535 VE

'Sand-island' (*v.* eg). There is water on two sides of Sandy
and 'the Greensand...on reappearing attains a considerable
elevation in the vicinity of Sandy' (VCH i. 10).

BEESTON [bi·sən]

Bistone 1086 DB, Hy 2 (Hy 3) *St Neot* 54, 1202 Ass, FF,
  1206 FF, 1227 Ass
Buistona Hy 2 (Hy 3) *St Neot* 54
Beston Hy 2 (Hy 3) *St Neot* 54, 1202 Ass, 1219 FF, 1223–4 Cl,
  1227 Ass, 1228, 1232 FF, 1236 FF, 1240 Ass, FF, 1247,
  1276, 1287 *Ass*, 1299 Ipm, 1301 Cl, 1312 AD vi, 1339 AD
  v, 1341 AD vi, 1346 FA, 1382 Cl, 1488 Ipm, Hy 8 BHRS
  ii. 116, 1553 BM
Beeston 1219, 1247 FF, 1287 *Ass*, 1342 Ipm, 1376 Cl, 1382
  IpmR, 1387 AD vi, 1428 FA, Hy 8 BHRS ii. 116
Bueston 1220 LS, 13th AD iii, vi
Boeston 1227 Ass
Beyston 1232 Cl, 1254 FF, Hy 8 BHRS ii. 114, 1549 Pat
Buston 1246 FF, 1247, 1276, 1287 *Ass*, 13th AD iii, vi, 1331
  QW, 1337 AD vi, 1488 Ipm, 1515, 1553 BM
Beuston, Bouston 1276 *Ass*
Bystone 1276 *Ass*

Bestun 1276 *Ass*
Boston(a) 13th AD vi, 1346 FA
Bieston 1362 AD iii
Bayston 1535 VE
Beeson Eliz ChancP

It is difficult to attain certainty with regard to this name, but one may suggest that it is from OE *byges-tūn*, 'farm of or in the bend of the river.' This would explain the variant ME forms and suits the topography as Beeston lies in a well-marked bend of the river Ivel or (alternatively) in the well-defined south-west angle of the parish of Sandy. The only difficulty in this interpretation is that it involves a rare type of place-name compound in which the significant first element is in the gen. case, but Ritter (155) has shown clearly that the existence of such in OE place-nomenclature, at least in the case of triple compounds, cannot be denied, and there is abundant confirmatory evidence in ME field-name material even for simple compounds, e.g. *Brochiseved* (1279), *Hulkescroft* (1304), *Parkesriding* (13th), *Sturteslowe* (1269), *Holmeshurne* in these two counties.

BEESTON DEAN

le Dene 13th AD vi, 1468 AD iii

Self-explanatory.

FEN FARM

mariscus 1212 Cur

GIRTFORD [gə·fəd]

Grutford 1247, 1276 *Ass*
Grotford 1247 *Ass*
Grutteford Hy 3 BM, 1276 *Ass*, 1338 Cl
Gretford 1291 Tax, 13th AD vi
Grotteford 13th AD vi
Gyrtford 13th AD vi, 1629 BM
Gretteford 13th AD vi
Gritteforde 1407 AD vi
Gyrford, Gurford 1634 BHRS i. 259, c. 1750 Bowen

'Gravel-ford' *v.* greot, ford.

GIRTFORD BRIDGE (6″)

pontem de Gretteford 13th (15th) *Newn* 9 *b*

THE HASELLS

Heyseles 1291 Tax, 1402 BHRS i. 102

'Buildings by the *hay* or enclosure' *v.* (ge)hæg, sele, or from heg, hence 'hay-buildings,' i.e. where it is stored. Cf. Newsells (Herts).

KINWICK (lost)

Chenemondewiche 1086 DB
Kenemundby 1212 Cur
Kenemundewich, Kenemundewyk 1220 *St Neot* 54, 1242 Fees 889
Kenendewyk 1242 Fees 889
Kynemundewyk, Kinemundewik 1252 FF, 1287 *Ass*, QW
Kynemundeswik 1276 *Ass*
Kylmundewyk 1332 Pat
Kynwyke 1535 VE
Kinwick Field 1804 *Award*[1]

'Cynemund-farm' rather than 'Cynemund's farm,' for there is a noteworthy absence of any forms with the genitival suffix. *v.* wic.

SANDY HEATH

bruera de Saunedey 1307 *Ass*

SEDDINGTON

Sudington 1306 Abbr (p)
Sodyngton 1307 *Ass* (p)

The history of this name is determined by that of a small group of Worcestershire place-names. Sinton in Grimley is *Suþtun* in BCS 386, but in the 13th cent. appears as *Suthintune*, *Sudinton*. It is in the south of the parish and is in contrast to Northington in the north. Leigh Sinton in Leigh has similar ME forms and is in the southern half of the parish. Sindon's Mill (and Sinton's End in old 1-in. O.S. map) in Suckley also has the same forms and is at the south end of the parish. Sud-

[1] *v.* BHRS v. 61–73.

dington in Ombersley, with similar forms, is in the south of the parish and is in contrast to Northampton in the same parish, and Sodington in Mamble has the same history. Seddington, whose early forms are identical with those just quoted, lies on the southern boundary of Sandy parish and it is clear that it has the same history. All alike must go back to *Sūðingatun*, farm of the dwellers in the south of the parish or manor[1]. *v.* ingtun. For the vowel cf. Diddington *infra* 254.

STRATFORD

> *Stratford* 1325 Ipm

This is a ford which must have carried the Roman road, of which mention is made under Stratton *supra* 102, across one of the feeders of the Ivel at this point.

SWADING HILL

It is difficult not to believe that this hill preserves the name of the family of Thos. de *Swathyng* who in 1346 (FA) held land in the not very distant Chawston and Colesden.

Sutton

SUTTON 84 G 13

> *Sudtone* 1086 DB
> *Sutton(e)* 1086 DB *et passim*
> *Sotton* 1284, 1316 FA, 1337 Fine
> *Sutton juxta Bicleswade* 1311 BM
> *Soutton* 1315, 1317 Ipm
> *Sutton Latymer* 1380 Cl

'South farm' in relation to Potton.

Tempsford

TEMPSFORD 84 E 11

> *Tæmeseford* 921 A (c. 950) ASC
> *Temesanford* 1010 E (c. 1200) ASC
> *Tamiseforde* 1086 DB, 1182 P, 1202, 1241 FF, 1242 Fees 869
> *Thameseford* 1219 FF
> *Tameseford* 1220 LS, 1227 Ass, 1246 FF

---

[1] For this explanation and all its details we are indebted to Mr F. T. S. Houghton, who has also shown that Siddington (Ch, Gl) and Sowton in Dunsford (D) have the same history.

Temeford 1227 Ass, 1245 Cl
Temiseford 1227 FF, Ass
Temeseford 1227 Ass, 1230 Cl, 1232, 1241, 1269 FF, 1294
   Ipm, 1337 Fine
Temseford 1228 FF, 1284 FA, 1349 AD vi
Tamesford 1240 Ass, 1323 Ipm
Themes(e)ford 1240 Ass, 1259 FF, 1380 Cl
Tamseford 1247 *Ass*
Temesford 1247 *Ass*, 1262 FF, 1276, 1287 *Ass*, 1316 FA, 1325
   Fine, 1331 Ch, 1346 FA, 1361 BM, 1363, 1380, 1395 Cl,
   1400 CS, 1428 FA, Eliz ChancP
Themisford 1276 *Ass*
Tamysford 1276 *Ass*
Tempisford 1526 LS
Temysford 1551 BM

The solution of this name is to be found in mutually com-
plementary statements made in the Anglo-Saxon Chronicle and
the *Historia Eliensis* (ed. Stewart, 139). In the former we hear
of the death of earl Toglos, a Danish chieftain, at Tempsford
in 921. In the Ely historian we are told that earl Toli (there is
little doubt about the identity of the persons or of the events)
was killed 'apud Tamensem fluvium.' It is clear from this that
the Ouse, or at any rate this stretch of it, must at an earlier date
have been known as the 'Thames.'

LAMBCOURT END
   Lambecotes 1287 *Ass* (p)
   Self-explanatory. *v.* cot.

MOSSBURY (Manor)
   This manor takes its name from the grant made in 1332 (Pat)
to John Morice, the present name being a corruption of
'Morice's bury' (*v.* burh). For the development cf. Mos-
borough (Db), earlier *Moresburh* (PN Db).

Wrestlingworth

WRESTLINGWORTH [reslɪŋwərθ] 85 F/G 1
   Wrastlingewrd c. 1150 BM
   Wrastlingeworde Hy 2 BM, 1198 FF (p)

Westlingewurda 1194 P
Wrastlingewurde 1195 P
Wrestligewrde 1197 FF
Wrestlingewurth 1212 BM
Wrastlingewrdh 1220 LS
Wrestlingeworth 1227 Ch
Wrastlingwrthe 1227 Ass
Wraxlingwurth 1232 Cl
Wrastlingworth 1232 Cl, 1234 Pat
Wrestlingwurth 1234 Cl

After this we have eight *a*- and five *e*- forms in the 13th cent.,
seventeen *a*- and three *e*- forms in the 14th, and *a*- forms are
still the commonest in the 15th cent. We also have

Wrestlyngworth or *Wrastlyngworth* 1335 Ipm
Warslyngwrth 1291 Tax
Warstlyngworth 1291 NI
Wraxlingworth 1434 IpmR

For this name Skeat aptly quotes the p.n. *Wræstles hyll* or
Wrestleshyll (Berks) found in BCS 789. These names point to
an OE pers. name *Wræstel, Wrǽstla*, a derivative of OE *wrǽste*
'delicate, noble' or of *wrǽstan* 'to twist,' but no such name is
known elsewhere. The whole name in OE must have been
Wræstlingaworþ, meaning 'enclosure of Wræstel's people.' *v.*
worð.

VII. MANSHEAD HUNDRED

Manesheue 1086 DB
Manesheuid 1175 P
Mannesheued 1176 P, 1227 Ass, 1247, 1287 *Ass*
Mansheve 1185 P
Manesheued 1202, 1240 Ass
Mauneshede (sic) FA *passim*

Manshead Hundred is one of the few Bedfordshire Hundreds
of which we know the approximate meeting place. Dr G. H.
Fowler (BHRS viii. 175) found in the Eversholt Enclosure
Award the following field-names: Great and Little Manshead
Closes, Manshead Path Close, Manshead Short Furlong and

Manshead Field, all close to the boundary brook between Eversholt and Tingrith parishes. The site to which these field-names refer is a long and low but well-defined hill, well fitted for the meeting-place of a considerable assembly in early times. It is clear that the site of the hundred meeting-place must have been close to this stream and the idea receives striking confirmation in the name Tingrith itself (*v. infra* 134). The brook by the meeting-place must have been known as *thing-rithe*, i.e. brook of assembly. Its position is now uncentral for the Hundred, but it was roughly at the centre before Stanbridge half-hundred was added. The name means what it says, viz. 'man's head.' For its interpretation cf. Swineshead *supra* 20.

Half-Hundred

STANBRIDGE

> *Stanburge* 1086 DB
> *Stanbrigge* 1207 *P*, 1284 FA
> *Stanburgh* 1227 Ass
> *Stanbrugge* 1316 FA

This half-hundred takes its name from Stanbridge *infra* 132 and is now absorbed in Manshead Hundred. In DB it included Studham, Eaton Bray, Totternhoe, Tilsworth and Nares Gladley in Heath and Reach. The meeting-place would almost certainly be at the ford rather than the present village.

Aspley Guise

ASPLEY GUISE [ɑˑspli] 95 A 4

> *Aepslea* 969 BCS 1229
> *Aspeleia* 1086 DB
> *Aspele(gh)* 1202 FF *et passim*
> *Aspeleye Gyse* 1363 Cl

'Aspen-tree clearing' *v.* æspe, æps and leah. Anselm de Gyse held the manor for one-twentieth part of a knight's fee in 1276 (VCH iii. 339).

BERRYLANE (Fm)

> *Bury Lane* 1530 *Dist. Prob. Reg. Northampton*, lib. iii. 35

The lane takes its name from the manor (*v.* burh).

FOTSEY (lost)

Foteseige 969 BCS 1229
Foteseye 1247 *Ass*
Fottesey 1247 *Ass*
Fodeseye 1292 Ch

Probably 'Foot's island.' There is no OE name *Fōt* on record, but Foston (Lei, Nf, L, Y) and Fosdyke (L) all go back to early forms in *Fotes-*. The distribution of this last group of names suggests that they contain a pers. name of Scandinavian origin, corresponding to the ON nickname *Fótr* (*v.* Lind, *Binamn* s.n.). This name does not seem very probable in a Beds charter of 969 but it is not wholly impossible, and the *Fot* of the present name is probably its OE cognate. The second element is *eg.*

RADWELL PIT (6″)

Rattle Pit in 18th cent. and so pronounced to-day (G. H. F.).

WENSDON HILL (6″)

Wendlesdun, Wændlesdun 969 BCS 1229

No personal name *Wændel* is found in the *Onomasticon*, but there can be no doubt of its existence in OE. Mr Bruce Dickins has kindly furnished us with the following parallels: Wendlebury (C, O), Wellingborough (Nth), Wendling (Nf), Wandsworth (Sr), Wansley (Nt), all of which go back to an OE *Wændel*. In addition to these he notes the unidentified *Uuendlesclif* in Gloucestershire (BCS 246) and *Wændlescumb* (KCD 1283) in Berkshire, *Wendlesbiri* (KCD 826) in Hertfordshire and, we may add, Winslow (Berks), formerly *Wendlescliue*. There are also OE compound pers. names *Wendelbeorht, Wendelburh* and (possibly) *Wendelgær* containing the same element. The name is common in OGer as *Wandil* or *Wendil* and is found in Norse mythology as *Vandill*, the name of a sea-king and of a giant (cf. Lind, *Dopnamn* s.n.). It is also found as the second element in ON *Ǫrvandill*, the name of a giant in Norse mythology whose toe was transformed into the star known as 'Orvandill's toe.' This name is found in OGer as *Aurivandala* and is clearly the same as OE *ēarendel*, a word for the dawn. These names and the occurrence of the word *Vandilsvé*, 'Vandill's shrine,' in *Helgakviða Hundingsbana* (ii, 34) unite in suggesting the possibility that Germanic *Vandilo* was originally the name of some mythological person, and that his name, like other names of divine beings, came to be used freely in the formation of personal-name compounds, which themselves could in turn undergo the usual shortening to pet-forms and give rise to such an OE pers. name as *Wændel*. The same element is probably to be found in OE *Wendelsæ*, OHG *Wentilseo*, both used of the Mediterranean, the latter being used also as a gloss for *oceanus*. *Wentilmeri* is a similar gloss for *oceanus*, and there is also an OHG *Wentilstein*. In this last group of names Mr Bruce Dickins suggests with much probability that we have this mythological name used with intensive force to denote something great, and compares the history of the element *eormen* in OE *eormengrund*, 'mighty deep,' originally the name of a divine being. The Vandals (Lat. *Vandali*, *Vandili*) must ultimately have derived their name from this word, but exactly in what sense is not clear.

Aspley Heath

v. Aspley Guise.

THE KNOLL (6″)

le Knol 1226–43 AD i

Self-explanatory. *v.* cnoll.

Battlesden

BATTLESDEN 95 C/D 5

Badelestone 1086 DB, 1227 Ass, 1276 *Ass*, 1315 Ipm

Badelesdone 1086 DB, 1152–8, 1155, 1160–5 NLC, 1179 BM, 1227 Ass, 1247 FF, 1287 *Ass*, 1292 Ch, 1302 FA, 1314 Cl, 1316 FA, 1361 Cl, 1397, 1399 AD i

Badelesdun c. 1160 (14th) Gest St Alb, 1220 LS, 13th *Dunst*

Badelesden 1227 Ass, 1276, 1287 *Ass*

Baddelesdon 1242 Fees 867

Badlesdon(e) 1247 FF, 1346 FA, 1348 Ipm, 1361 Cl

Batlesden 1254 FF, 1287 *Ass*, 1428 FA

Battlisden 1257 Ch

Baddesdon 1284 FA

Badlesden 1287 *Ass*

Baddelesden 1287 *Ass*
Batelesdon 1347 Cl, 1348 Ipm
Badleston 1348 Ipm, 1391 IpmR
Bat(t)elesden 1389 IpmR, 1489 Ipm
Badlysdene 1390–2 CS
Badlysden 1526 LS
Batellesden 1535 VE

The existence of an OE pers. name *Badel(a)*, a diminutive of the recorded name *Bada*, is made certain by the p.n. *Badelanbroc* (KCD 715) and by Badlingham (C). The present name, therefore, means 'Badel's hill,' *v.* dun. The strong form *Badel* occurs again in Badlesmere (K). Sound dissimilation has done its best with this name, leading sometimes to a suffix *-ton* and later to a prefix *Batel-*.

Billington

BILLINGTON 95 E 4
Billendon 1196 P
Bilindon, Bylindon 1196 FF, 1349 BM
Billesdon 1202 PR
Bilendon, Bylendon 1227 Ass, 1247 *Ass* (p), 1249 FF, 1276 *Ass*, 1297 *SR*
Billingdon, Billyngdon 1276 Ass, 1491 Ipm
Bellendon 1276 *Ass*
North Billingdone 1287 *Ass* (p)
Belyngdone 1393 BM, 1394 AD vi
Bellingdon, Bellyngdon 1395 IpmR, 1491 Ipm
Billington 1798 Jury

'Billa's hill' *v.* dun. The later *-ing-* forms may be a corruption of earlier *Billandun* or they may be due to a new form in *-ingdun* having arisen side by side with those in *-andun*. For such formations *v.* ingtun.

Chalgrave

CHALGRAVE 95 D 6
Cealhgræfan 926 (12th) BCS 659
Celgraue 1086 DB
Chealgraue 1163–87 *Merton* 809

Chalgraue 1173 P, 1276 *Ass*, 1284 FA, 1291 NI, 1302, 1316, 1346 FA, 1366 Cl *et passim*
Calgraua a. 1177 (13th *Dunst*)
Chalkgrave 1220 LS, 1308 Cl
Chaugrave 1220 (13th) *Dunst*, 1242 Fees 887, 1247 *Ass*, 1257 FF, 1276 *Ass*, 1286 Dunst
Chagrave 1286 Dunst
Chalfgrave 1346 FA

There can be no doubt that Skeat's conjecture that the first element in this name is OE cealc is right, for Mr Gurney in his admirable study of the bounds of Chalgrave (BHRS v. 163 ff.) shows that, somewhat unexpectedly, chalk is found here and that the shepherds in driving their stakes frequently strike hard chalk. The second element, as almost always in these *grave*-names, is not quite certain. If we may judge by the form of the name of the other ground assigned in the charter, viz. Tebworth *infra* 118, we must take *-græfan* to be in the dative. If so it could be either the dat. pl. of græf and the name be interpreted 'chalk-pits,' referring to some small diggings for chalk made here, or the dat. sg. of græfe. 'Chalk-thicket' seems however to be a somewhat unlikely name.

HILL FM
   la Hulle 1297 *SR* (p)
   Self-explanatory.

KIMBERWELL (local)
   (*to*) *cynburge wellan* 926 (c. 1200) BCS 659
   This well clearly takes its name from one *Cyneburh*, who may perhaps be identified with St Cyneburh, the daughter of Penda of Mercia, to whom it may have been popularly dedicated. (See further BHRS v. 168.)[1] She has left her name also in the Coney-burrow Way near Peterborough, while in early days the neighbouring Castor (Nth) was *Cyneburgecæstre* (BCS 871), and the church is still dedicated to her (*ex inf.* Mr Bruce Dickins).

[1] Mr St Clair Baddeley informs us that outside the former South Gate of Gloucester she was held in high veneration and a well was tragically associated with her death in one of the miracle stories of her there. She had a chapel there. The name in Gloucester became *Kimbrose, Kimbrow*.

L

THE OLD BROOK (local)

þane ealdan broc 926 (c. 1200) BCS 659

v. BHRS v. 168.

TEBWORTH

Teobbanwyrþe 962 (c. 1200) BCS 659
T(h)ebbeworth 1227 Ass, 1276 *Ass* (p), 1286 Dunst, 1365 AD i
Tebbewurth 1227 Ass, 1247 *Ass*
T(h)eburthe 1286 Dunst
T(h)ebworthe 1286 Dunst, 1308 Cl

'Teobba's enclosure' *v.* worþ. The pers. name *Teobba* is not on independent record but it is clearly a pet-form for some such OE name as *þeodbeald, þeodbeorht* or *þeodburh* (a woman's name).

WINGFIELD

Winfeld, Wynfeld c. 1200 (13th) *Dunst*, 1225 FF, 1227 Ass, 1247 *Ass*, 1276 *Ass* (p), 1286 Dunst
Wintfeld 13th *Dunst*
Winefeld, Wynefeld 1249 (13th) *Dunst*, 1286 Dunst, 1317 AD i
Wynchefeld 1276 *Ass* (p)
Wyndeselde (sic) 1535 VE
Winfield 1675 Ogilby

The early forms are indecisive. *Wintfeld, Wynchefeld*, here recorded for the first time, make Skeat's suggestion of a pers. name *Wina* unlikely. Possibly the first element is the word *wince*, used of a nook or corner, the likelihood of which was demonstrated in PN Bk 203. Wingfield does stand at the head of a little valley. If the last theory is the correct one the *ch* must early have been lost between the *n* and the *f*. It is clear in any case that the *ng* for *n* is a late corruption.

Professor Ekwall suggests that this name may be connected with the OE pers. name *Winta*. A strong genitive form *Wintesfeld*, side by side with the correct *Wintanfeld*, would, with common development of *ts* to *ch*, explain *Wynchefeld*.

Husborne Crawley

HUSBORNE CRAWLEY [*olim* hʌzbənd] 95 A 8

(of) *Hysseburnan* 969 BCS 1229
Husseburn(e) c. 1200 (13th) *Dunst*, 1220 LS, 1227, 1247 *Ass*, 1249 FF, E 1 *KF*, 1274 Cl

Hesseburn 1220 (13th) *Dunst*
Hisseburn c. 1220 WellsL
Husburne, Husborne 1250 FF, 1428 FA
Husseburn (et) Crawele, Husseburn Crouleye 1276 *Ass*
Hussebourne 1291 NI
Husbond Crawley 1535 VE, 1548 Pat
Husband Crauley 1552 Inv
Husband 1615 Cai

Husborne Crawley takes its name from the union of two distinct places called Husborne and Crawley. The stream from which *Husborne* takes its name has in OE a name identical with that for Hurstbourne (Ha), which is found as *Hysseburnan* in BCS 553. Both names have undergone later corruption and the local names of Hurstbourne Tarrant and Hurstbourne Priors— Uphusband and Downhusband—show the same colloquial pronunciation that we find in the Bedfordshire p.n. The history of this stream-name is obscure. There is an OE word *hysse*, more commonly *hyse*, used to denote a young man, a warrior, and it is just possible that *Hysseburna* means 'young men's stream,' but if that were the case we should have expected *hyssa burna*, with *hysse* in the gen. pl., as it seems to be found in the only other possible compound that has been noted, viz. *hyssa pol* in BCS 595 referring to a Wiltshire place. The fact that in both cases the word is found with a word for water suggests however that it may possibly be a plant-name. Cf. *hisses* quoted in B.T. Supplement (s.v. *hyse*) as a gloss for *pampinos*. No certainty can be attained with regard to this name.

CRAWLEY

Crawelai, Crauelai 1086 DB
Craule(e) 1205–50 (13th) *Dunst*, 1221 FF, 1242 Fees 890,
    1287 *Ass*, 1291 NI, 1346 FA, 1368 Cl, 1428 FA
Crawele, Crauele 1227, 1240 Ass, 1249 FF, 1276 *Ass*, 13th
    AD v, 1316 FA
Crouleg, Crouleye 1247, 1276 *Ass*
Crowley 1530 LP
Crawley 1535 VE

'Crow clearing' (*v.* crawe, leah), with the same hesitation between a development to *Crowley* or to *Crawley* that was discussed under North Crawley in PN Bk 34.

Dunstable

DUNSTABLE 95 E/F 6/7

Dunestap(e)le (dat.) 1123 E (12th) ASC, 1130 P *et passim* to 1400
Dunistapla 1173 P
Dunstaple 1189 (1227) Ch, 1199 FF *et passim* to 1766 J
Donestaple 1202 Ass *et passim* to c. 1400
Dunnestaplia 1224 Bract
Donestable 1287 *Ass*
Dunstable 1287 QW
Dounstaple 1307 *Ass*

Perhaps 'hill post or pillar' (*v.* dun, stapol), or 'pillar of the hill,' the reference being to some such erection at the spot where on open and high ground the Watling Street and the Icknield Way intersect. The situation of Dunstable, in a gap in the chalk hills of south Bedfordshire, makes derivation from dun, 'down,' superficially probable. On the other hand, the numerous early forms with *e* between the two elements of the name suggest that the first element is a pers. name *Dun(n)a*. To an OE *Dunnan stapol*, 'Dunna's pillar,' there are exact parallels in Barnstaple (D) and Barstable (Ess), each of which represents an OE *Beardan stapol*. The medieval legend that Dunstable was so called from one *Dunning*, who had as a thief frequented the site of the future town, receives no support from the forms which are collected here (Dugd. vi. 239). It shows, nevertheless, that the men of the neighbourhood did not believe that the name was derived from the down above Dunstable, and it may possibly preserve a distorted memory of the *Dunna* who left his name to the place and, as suggested by Mr Bruce Dickins, the *i* of the 1173 form may be from a form *Dunning-stapol* with *ing* for gen. *an* (*v.* ing).

GOLDENLOW (lost)

Goldenelowe 1286 Dunst
Gyldenelowe 1292 Fine

'Golden hill,' from OE hlaw, with alternative forms of the first element from OE *gylden* or ME *golden*. The source of the name is to be found in the mention in the *Annals of Dunstable* (363) of an enquiry which took place in the reign of Edward i with reference to certain treasure trove found at *Goldenelowe* in the days of Henry iii.

HOLLIWICK ST (lost)

le Hallewick 13th AD i
le Hallewyke 1317 AD i, 1339 AD vi

This lost street ran parallel to West Street on its southern side and the VCH (iii. 350) gives a 17th cent. form *Holliwick*. This would point to OE halig wic, 'holy *wic*,' perhaps so called because the farm from which it took its name was monastic property, though we have no knowledge of such possession. One would have expected however ME forms with a single *l* if this were the history of the name. Cf. *Halwic*, *Haliwyk* for Holywick (PN Bk 179). One might, with Professor Ekwall, take this to be OE heallwic, i.e. 'the dairy-farm belonging to a hall,' but this leaves the modern form unexplained.

KINGSBURY HOUSE (local)

This is the last trace of the 'domus et gardinium regis in Dunstaple,' which goes back to 1204 (ChR). *v.* burh.

Eaton Bray

EATON BRAY 95 F 5

Eiton(a) 1086 DB, 1130 P *et passim* to 1286 Dunst
Eitun 1156 P
Æiton 1159 P
Etton 1164 P
Ehton 1166 P
Eyton 1220 LS, 1240 Ass, 1247 *Ass et passim* to 1491
Eton 1241 Cl, 1330 Ipm, 1338 Cl, 1490 Ipm

'River-farm' (*v.* eg). Eaton Bray is a district with numerous small streams and eg must be used in the wider sense of land well watered rather than of an actual island, as is fairly common in place-names. The feudal addition belongs to the 15th cent. when (in 1490) the manor was granted to Sir Reginald Bray (cf. VCH iii. 371).

Eggington

EGGINGTON 95 E 5

Ekendon 1195 FF, 1234(13th) *Dunst*(p), 1276 *Ass*(p), 1297 *SR*
Ekyndon, *Ekindon* 1276 *Ass*, 1341 BHRS viii. 26
Egynton 1304 Pat

'Oak-grown hill' with the same adj. *æcen*, that was found in the lost Ekeney (PN Bk 37). Such a development of medial *k* to *g* is fairly common and is fully illustrated in PN Bk 83, *s.n.* Bragenham.

CLIPSTONE

> *Clipeston* R i Cur
> *Clapston* 1195 P
> *Clipston, Clypston* 1196 P, 1247, 1276 *Ass*, 1287 *Ass* (p)

'Clip's farm.' The pers. name *Clip* is found once as the name of a 10th cent. moneyer, but the distribution of the place-names in which it is certainly found, viz. Clippesby (Nf), Clipston (Nth, Nt), Clipstone (Nf), Clipsham (R) makes it certain that the name is Scandinavian rather than English and that Björkman (ZAN 55) was right in associating it with the ON nickname *Klyppr*, which was used of some one with an awkward stumpy figure (cf. Lind, *Binamn* s.n.).

EDE WAY (local)

> *ðiodweg* 926 (c. 1200) BCS 659

The term *ðiodweg*, a compound of *ðeod*, 'nation, people' and *weg*, is fairly common in the OE charters, and like the similar compound *ðeodherepæð* is the equivalent of what in the Latin versions of the charters is called a *via publica*. Mr Gurney (BHRS v. 169) identifies it with a wide and open green-lane on the low (greensand) ridge. To quote his own words: 'It possesses several names in different parts of its course....At Wingfield in Chalgrave it serves as the present hard road and thence continues as a footpath to Kateshill on Watling St.... At Eggington it reappears as an unusually broad green-way, under the name of the Ede Way.' He then shows how this road ultimately makes its way to the Ouzel at *Yttingaford* (cf. PN Bk 81), locally Tiddingford Hill. One interesting point, however, he misses in his masterly treatment of the charter. The *Ede Way* represents not only the course but the actual name of the old *ðiodweg*. This last word would develop to *Thedewey* in ME. Later, by a common process of misdivision, this was taken to be for *The Edewey*, and hence the modern name.

Eversholt

EVERSHOLT [evəsɔ·l] 95 B 6

Eures(h)ot 1086 DB
Ewersolt (sic) 1162–79 HMC, Var iv
Euresholt 1185 P
Euereshold 1202 Ass
Evereshout 1202 FF, 13th *Dunst*
Everesholt 1219 FF *et passim*
Euersholt 1221 FF *et passim*
Heueresholt 1240 Ass *et sæpe*
Eversolt 1291 NI
Eversoll 1518 *Award*
Evershold 1526 LS
Eversoult 1552 Inv
Eversal(l) Eliz ChancP, 1722 NQ ii
Evershall 1605 NQ ii

'Boar's wood' *v.* eofor, holt.

FROXFIELD

Frogfield 1826 B

This p.n. almost certainly has the same history as Froxfield (Ha, W) and means 'feld of the frogs.' *v.* forsc.

HAY WOOD

Heywode 1287 *Ass* (p)

'Enclosed wood' *v.* (ge)hæg, wudu.

WAKE'S END

Eversholt Wakes 1826 B

The last relic of a Wake Manor in Eversholt (cf. Ch 1313).

Harlington

HARLINGTON 95 C 7

Herlingdon, Herlyngdon 1086 DB, 1227 Ass *et passim* to 1428 FA
Erlingedun 1181–90 (13th) *Dunst*
Herlingedune 1181–90 (13th) *Dunst*, 1286 Dunst
Herlingedon 1189 P, 1223 FF, 1227, 1240 Ass

Harlingedon 1220 LS
Herlincdon 1223 (13th) *Dunst*
Herlington 1240 Ass, 1423 IpmR
Herlindon, Herlyndon 1278 Cl, 1287 *Ass*
Harlyngdon, Harlingdon 1489 Ipm, 1548 Pat, 1552 Inv
Harlyngton 1492 Ipm, 1526 LS

'Hill of Herela's people' *v.* inga. The *Herelingas* (*Widsith* 112) or Harlung brothers, nephews of Ermanaric the Goth, were among the most famous of the Germanic heroes. Apart from place-names however we have no evidence for the use of the name *Herel*(*a*) in England. In place-names it is found in Harling and Harleston (Nf), Harlethorpe (Y), Harlesthorpe (Db), Harlton (C), Harlington (Y). It is clear that the name must have been passing out of use soon after the period of the English settlement, and its distribution shows it to be definitely Anglian.

GOSWELL END[1] (6″)

Goseland, Gosefeld 1434 BHRS viii. 91
Gosly End 1677 ib.

It is clear that there has been a good deal of corruption going on here. The original form would seem to have been *gorstig-land*, 'gorse-grown country.' Cf. *Gostilaunde* in Eaton Socon (1226).

Heath and Reach

HEATH 95 D 4

la Hethe 1276 *Ass*, 1287 *Ass* (p)
Hethe 1297 *SR*
Hetheredge c. 1750 Bowen
Heathanreach 1785, 1791 Jury

v. hæð.

NARES GLADLEY

Gledelai 1086 DB
Gledele(*g*) 1131–41 (13th) *Dunst*, 1247 *Ass*, 1286 Dunst, 1331 QW, 1332 Ch, 1347 Ipm, 1379 Cl
Gledeleia 1176 P

---

[1] Other intermediate forms *Gosiland, Gostilaunde, Gosling End* are given by Mr J. H. Blundell in the article from which the first references are taken.

Gladelaia 1176 P (Chanc Roll)
Gladelea 1177 P
Gledley or *Gledeley* 1499 Ipm

There is no evidence in the charters for a topographical use of OE *glæd*, 'bright,' apart from Glatton *infra* 187. It is more probable that the present name contains a pers. name *Glǣda*, from OE *glǣd*, 'bright,' by the side of *glæd*, 'glad.' Cf. Ekwall in *Angl. Beiblatt* xxxiii, 67. *Nares* is presumably the name of some owner or tenant, but remains an unsolved mystery.

HATCH FARM (6")

la Hacche 1287 *Ass* (p)

v. hæcc. To judge by the situation the reference was rather to a gate than to a sluice or weir.

KING'S WOOD

Kyngeswode 1307 *Ass*

This probably gained its name when Heath and Reach was part of the royal manor of Leighton Buzzard.

REACH

Reche 1216–31 (c. 1350) Rams, 1276 *Ass*, 1287 *Ass* (p), 1321 BHRS iii. 24
Rache 1276 *Ass*
Rach 1669 NQ i

That the word *reche* could in ME be used in a topographical sense seems to be clear from this place and from Reach (C). It is difficult however at first to see what they have in common. Skeat says of the latter place that old maps show that it stood at the very verge of the waters of the fen-lands, on a round projection of the old shore. Reach (Beds) lies on rising ground in a shallow valley, the village lying along the road which runs up the valley. Skeat takes the word to mean a *reach* or extension of the land in the Cambridgeshire p.n., but there is really no evidence for such a use of 'reach' in Middle English. *Reach* in the sense 'that which reaches or stretches' is no older than the 16th cent. What Skeat did not notice, however, is that Reach lies just at the end of the Devil's Dyke. That dyke indeed has served its purpose when it reaches the edge of the Fens; in early days it must however have formed a natural path of approach to Reach, which to this day is singularly inaccessible.

There is an ON *rák*, 'stripe, streak,' which is the source of the English *rake*, 'way, path, narrow path up a cleft or ravine,' and there is a cognate *rack* from a different ablaut-grade, meaning 'narrow path or track,' of such wide distribution in the south and west of England that it must be of native English rather than Scandinavian origin (*v. rake*, sb. 2 and *rack*, sb. 1 in NED and *rack* in EDD). Possibly we may note also Wiltshire *rake*, 'row of houses,' EDD. Corresponding to either of these words there may in OE have been an OE *rǣc* or *ræcc*, *jo*-stems with consequent palatalisation of the *c*. Either of these would account for ME *reche* or *rache* such as we find in the early forms of Reach. The possibility of such a derivative receives some measure of confirmation from such a noun as *rache, ratch* (rarely *reach, race*) used in English of a white streak on a horse's face, in which we note the same sense of a narrow line (NED s.n. *rache*, sb. 2, *race*, sb. 5). It may well be, if this etymology is correct, that the Bedfordshire place was so called because it lay along a steep narrow road running up a valley, and the Cambridgeshire place from the Devil's Dyke itself which was used as a *reach* or path.

Hockliffe

HOCKLIFFE 95 D 5

    Hocganclif 1015 Thorpe 561

    Hocheleia 1086 DB

    Hoccliue, -lyue 1185 P, 12th HMC, Var iv, 1220 LS, 1228 FF, 1240 Ass, 1242 Fees 885, 1276, 1287 *Ass*, 1291 NI, 1302 FA, 1388 Cl

    Occliue 1227 Ass, c. 1370 *Linc*

    Houclive 1247 *Ass*

    Hoklyue, Hoclyue 1247, 1276 *Ass*

    Hocclyve al. *Hoclyve* 1324 Ipm

    Oclyve 1346 FA

    Hocklyve 1428 FA

    Occleue c. 1460 *Linc*

    Hockley 1576 Saxton, 1633 NQ i, 1675 Ogilby (alias *Hockley in the Hole*)

'Hocga's cliff,' the church, which probably forms the nucleus of the original parish, standing on a steepish spur of land to the north-west of the present village. The latter well deserves its popular name of 'Hockley-in-the-Hole' for it lies along Watling Street in a well-marked depression, the road to the northwest rising from the village some 117 ft. in half a mile. It was famous for its robberies and must be distinguished from Hockley-in-the-Hole (of equally bad fame) in the Fleet valley in Clerkenwell.

Loss of final *f* is fairly common in p.n., cf. [jakli] and [kunzli], the local pronunciations of Aycliffe and Coniscliffe (Du), and *v.* PN NbDu 265. For the personal-name *Hocg(a)*, cf. Hoggeston (Bk) and Hoxton (Mx). The poet (H)occleve must have derived his name from this place.

Shenley Hill (6″)

Sinelehe 1276 *Ass* (p)
Shenle 1276 *Ass* (p)

'Bright clearing' *v.* sciene, leah. This is fairly common as a p.n.

Holcot

Holcot [hʌkət] 84 J 4

Holacotan (g. pl.) 969 BCS 1229
Holecote, Holekote 1086 DB *et passim* to 1428 FA
Hulecote 1240 Ass
Holcote 1315 Ipm *et passim*
Hulcote 1461 IpmR
Hulcott Eliz ChancP
Howcott al. *Hulcott* Eliz ChancP

Holcot lies in a small valley and the meaning of the name would seem to be 'cottages in the hollow' (*v.* holh), but it is a little difficult to see just how the OE compound is formed. It may be from the gen. pl. **hola**, hence 'cottages of the hollows' or the form may be a mistake for *holan cotan*, 'hollow cottages,' if that is a likely phrase. An OE noun *hola*, suggested by Skeat, is of very doubtful authenticity.

Brook Farm

Broke 1287 *Ass* (p)
Self-explanatory.

Houghton Regis

HOUGHTON REGIS 95 E 7

Houstone 1086 DB
Hohtun, Hohton 1156 P, 1227 Ass
Hocton 1157 BM, 1169 P, 1239 Ch, 1247 *Ass*, 1249 FF
Hochtun, Hochton 1158 P, 1227 Ass
Houcton 1169 P
Hoghton 1220 LS, 1247, 1287 *Ass* (*extra Dunstapel*)
Houtun, Houton Hy 3 (1317) Ch, 1242 Fees 869, 1284 FA
Kyngeshouton 1287 *Ass*
Houghton Regis 1353 Ipm

'Farm on the hoh or spur of land.' *Regis* because already in DB it was a royal manor.

BIDWELL

Budewelle 13th *Dunst*, 1228 FF, 1279 RH, 1297 *SR* (p)
Bedewelle 1279 RH

'Byda's spring' *v.* wielle and cf. Biddenham *supra* 26.

BREWERSHILL (Fm)

la Bruere 1307 *Ass*

If this identification is correct the name should really be 'Brewer Hill' and the first element interpreted as the equivalent of ModFr *bruyère* and used of heath-land. Cf. Sandy Heath *supra* 109.

BURY SPINNEY (6″)

cf. *Buridene* 1295 FF

The spinney lies in a valley (*v.* denu) which took its name from the manor of Thorn or Thornbury (*v. infra* 129).

CALCUTT (Fm)

Caldecote 1224 FF, 1247 *Ass et passim*
Caudecote 1247 *Ass*

'Cold cottages,' probably from their exposed situation. *v.* cald, cot.

PUDDLE HILL

Pudele 1274 (13th) *Dunst*
Pod(d)ele 1276 *Ass*
Podele 1304 AD iii

'Puda's clearing' *v.* leah and cf. Podington *supra* 37.

SEWELL

Sewelle 1086 DB, 1260 FF
Seuewell 1193 P, 1227 Ass, 1228 FF, 1286 Dunst
Sewell 1247 *Ass*, 1260 FF
Siwell, Sywell 1287 *Ass* (p)
Seywell 1287 *Ass* (p), 1766 J

This is a difficult name. The only suggestion that can be offered is that the first element is a pers. name *Seofa*, not on record, but of which we have a diminutive in Seacourt (Berks), *Seofocanwyrð* in BCS 1002. The same name seems to occur in Seawell (Nth). *v.* wielle.

THORN

la Thorn 1225 FF (p), 1316 AD i
Thornbury 1324 Pat, 1427 IpmR

Self-explanatory. The *bury* is manorial, *v.* burh.

Leighton Buzzard

LEIGHTON BUZZARD 95 D/E 4

Lestone 1086 DB
Le(c)htone 1164, 1169 P, 1227, 1240 Ass
Lechtone 1173 P, 1202 *P*
Leocton 1177 P
Lectune 1194 Cur(P)
Lecton 1195 P, 1196 FF, 1202 Ass, 1228 Cl, 1253 AD iii
Leiton, Leyton 1206 FF, 1227 Ass
Leghton 1247, 1287 *Ass* (*Busard*)
Lachton 1264 Pat

Further forms of Leighton are without interest. For the Buzzard we may note

Bussard 1287 QW, 1321 Cl, 1656 BM
Busard 1287 *Ass*, 1297 *SR*, 1307 Cl, 1316 FA, 1326 Cl, 1349
    BM, 1446 HMC Var iv, 1477 AD i

Bousard 1331 QW
Bosard 1355 Cl, 1447 IpmR, 1499 Ipm
Bussarde al. *Budeserte* Eliz ChancP
Beaudesert 1643 HMC v
Beudesert al. *Budezard* 1646 NQ iii
Buzard 1649 HMC v
Beuzard 1655 NQ i

'Kitchen-garden' is the meaning of OE leactun from which this p.n. comes. The Buzzard addition has never been satisfactorily explained. *Busard* is the name of a well-known French family, but their nearest property so far as we know was at Knotting in the north of the county. The *Beaudesert* of the later forms is clearly an antiquarian invention due to a determination to find some meaning for the name.

COLLICK (local) 'meadow and runnel'

Cochlake 1324 BHRS viii. 42

'Cock-stream,' because haunted by such, cf. *cocbroc* BCS 675.

CORBETSHILL (Fm) (6″)

This takes its name from the Corbet family (VCH iii. 405).

GROVEBURY

Grava 1195 FF (p)
la Grave 1245 Cl
le Grove 1310 Ch
Grovesbury 1363 Cl
Grovebury, Grovebyry 1387 Cl, 1390–2 CS

Self-explanatory. The *bury* is manorial, *v.* burh.

Milton Bryant

MILTON BRYANT 95 C 5

Mildentone 1086 DB
Middelton 1086 DB, 1220 LS, 1227 Ass
Middleton 1227 Ass
Midelton Brian E 1 KF
Mylton Bryan 1489 Ipm
Mylton Bryon 1535 VE

'Middle farm,' but why so called is uncertain. It does stand
about half-way between Woburn and Toddington. The *Brian*
tenancy of the manor goes back to the reign of Henry ii (cf.
VCH iii. 418).

Potsgrove

POTSGROVE   95 C 4/5

> *Potesgraue*   1086 DB, Hy 2 (1260) Ch, 1202 Ass, 1219 FF,
>   1242 Fees 894,   1247 *Ass*,   1284 FA,   1285 Ipm,   1291 NI,
>   1308 Cl, 1316, 1346 FA, 1348 Ipm, 1390–2 CS, 1428 FA
> *Pottesgrave* Hy 3 (1315) Ch, 1247, 1287 *Ass*, 1302 FA, 1489,
>   1499, 1504 Ipm
> *Pattesgrave* 1220 LS, 1276 *Ass*
> *Portesgrave* 1242 Fees 868, 1428 FA
> *Podesgraua* 1247 *Ass*, 1315 Ipm, 1373 IpmR
> *Potsgraue* 1254 FF
> *Potegraue* E 1 *KF*, 1276 *Ass*

'Pot(t)'s grove' *v.* grafa. A pers. name *Pott* is not on record
in OE but Skeat quotes *Pottingtun* from an OE charter and we
seem to have a diminutive of this pers. name in *Potteles treow*
(BCS 924).

Salford

SALFORD [sa·fəd]   84 J 4

> *Saleford* 1086 DB, 1156 P *et passim* to 1302 FA
> *Seleford* 1198 Fees 9
> *Salleford* 1220 LS
> *Salford* 1247 *Ass et passim*
> *Shalford* 1276 *Ass*
> *Sawford* 1610 Speed, 17th BHRS viii. 157
> *Salford* al. *Sawford* c. 1650 *Linc*
> *Sarford* c. 1750 *Strip*

'Willow-ford' *v.* sealh, ford.

WHITSUNDOLES (Fm)

Dr G. H. Fowler tells us that in 1595 this land was called
Lott Meades, marked as in 'x lottes' or 'x doales,' and seems to
have been allotted annually in ten shares to certain tenements.
He suggests that possibly the allotment was made at Whit-
suntide.

Stanbridge

STANBRIDGE 95 C 5

Stanbru(g)g(e) 1165, 1175 P, 1220 LS, 1242 Fees 867, 1276,
　　1287 *Ass*, 1316 FA, 1326 Cl, 1405 AD i
Stanbregge 1196 FF(P)
Stanbrig(g)e, Stanbrygge 1202, 1227, 1240 Ass, 1247 *Ass*, Cl,
　　1284 FA, 1353 Cl, 1405 AD i, 1489, 1497 Ipm
Staunbrig(g) 1227 Ass, 1247 *Ass*
Stanburgh 1227 Ass
Stantbrig 1240 Ass
Stanbrich 1276 *Ass*
Stambrug 1292 Ipm
Standbridge 1785 Jury

'Stone-bridge' *v.* stan, brycg. The village must have taken
its name from the bridge at Stanbridgeford, a good mile to the
south-east.

THE SLOUGH (6″)

la Slo 1297 *SR* (p)
la Sclo 1314 AD ii (p)

Self-explanatory. Cf. Slough (PN Bk 193, 243), all from OE
slōh.

Studham

STUDHAM 95 G 7

Stodham 1053–66 (c. 1250) KCD 945 *et passim* to 1526 LS
Estodham 1086 DB
Stodeham 1286 Dunst, 1310 Fine, 1316 Cl, 1491 Ipm
Studham 1329 Cl
Studham or *Stadham* c. 1750 Bowen

'Stud-homestead' *v.* stod, ham.

BARWYTHE

Bereworde 1086 DB
Baresworth 1200–1 Cur
Bareswurth 1200 FF
Barewurth 1219 FF, 1236 (13th) *Dunst*, 1262 FF, 1286 Dunst
Barewrthe 1220–40 (13th) *Dunst*, 1220 FF

Bareworthe 1259 FF, 1276 RH, 1376 Cl
Barworth 1806 Lysons

This may contain an OE pers. name *Bǣre* (with weak form *Bǣra*) which is on record in the 8th cent. This seems to be found in Barrington (C) and in Husbands Bosworth (Lei), which is probably a duplicate of the present name. A diminutive of it occurs in Barlings (L). Hence 'Bære's enclosure,' *v.* worþ.

BUCKLESHORE (lost)

Bukeleshore 1200 FF, 1286 Dunst
Buclesore 1201 Cur, 1220 FF, 1220–40 (13th) *Dunst*
Buclesoure 1202 Ass
Bokelesore 1203 Abbr, Cur
Bokeleswrth 1203 Cur
Buckellesore 1220–40 (13th) *Dunst*
Buckesore 1240 Ass
Bukkelesore 1247 *Ass*
Bokeleshore 1286 Dunst
Buklesore 1309 Ch

'Buccel's bank' *v.* ofer. The name is not on record but is a regular diminutive formation from *Bucca* and is clearly found in Bucklesham (Sf).

HILL FARM

la Hulle 1198 FF (p)

Self-explanatory.

Tilsworth

TILSWORTH 95 E 5

Pileworde 1086 DB
Thuleswrthe 1202 FF
Tillesworde 1219 FF
Tilesworth 1227 Ass
Tuillesworth 1227 Ass
Tullesworth 1240 Ass, 1247 *Ass*
Tiueleswrth 1247 *Ass*
Tilleswurth 1247 *Ass*
Tyuelesworth 1247 *Ass*
Tulesworth E 1 *KF*, 1297 *SR*, 1302, 1316, 1346 FA

M

Tulleworth 1276 *Ass*
Tullesworth 1276 *Ass*, 1284 FA, 1291 NI
Tulisworth 1276 *Ass*
Teulesworth 1276 *Ass* (p)
Thulesworth 1276 *Ass*, 1287 *Ass* (p)
Tyllesworth, Tillesworth 1276 *Ass* (p), 1293 Ch, 1367, 1381 Cl,
  1405 BM, 1407 AD i, vi, 1489, 1497 Ipm
Tyllisworth, Tyllysworth 1287 *Ass* (p), 1428 FA
Tylysworth 1390–2 CS
Tildesworth c. 1440 *Linc* (six times)

The forms set forth above show that Skeat's explanation of the first element as from a pers. name *Tugol* or *Tull* is inadequate. *Tulla* is known and a mutated form *Tyll(a)* would go rather further, but it would still leave unexplained those forms which seem to show the existence of a medial *v* (probably from OE *f*) in the early history of the name. Professor Ekwall calls our attention to the existence of a 12th cent. name *Thuf* (Farrer, *Early Yorkshire Charters*, no. 899). *Þȳfel* would be a regular diminutive form from this, and so Tilsworth would be 'Þyfel's worð.' If that is the case the initial *p* of the DB form is due to the common confusion of OE *p* and *þ*. This pers. name would be identical with the significant word *þȳfel*, 'thicket,' itself a diminutive of *þūf*, used in the slightly different sense of 'tuft.' It can hardly be that word itself, for worð is not very likely to be compounded with the gen. of a significant word.

BLACKGROVE WOOD (6″)

Blackegrave 1331 QW

Self-explanatory.

Tingrith

TINGRITH 95 B 6

Tingrei 1086 DB
Tyngri, Tingri, Tyngry, Tingry(e) 1220 LS, 1227 Ass, 1247
  Ass et passim to 1365 Cl
Tingerithe 1247 Ass
Tyngrithe, Tyngryth, Tingrith 1276 Ass, 1331 Ipm, 1390–2
  CS, 1509 AD v

Tyngre 1346, 1428 FA, 1489 Ipm
Tingryffe 15th HMC Var iv, 1605 NQ ii
Tyngreve 1504 Ipm
Tyngriff, Tyngryff, Tingrif 1526 LS, 1535 VE, 1566 BM
Tingreth 1598 D, Eliz ChancP
Tyngrave 1626 HMC iv, 1646 NQ iii
Tyngeriff c. 1690 *Strip* (Hail Weston)

This is clearly from OE þing and riŏ, the whole name meaning 'assembly-brook.' As we have seen above (p. 113) the name is peculiarly apt as the meeting-place of the Hundred of Manshead was by this brook. For the change of the suffix to *-riff*, Wyld gives interesting parallels, *Lambeff* for Lambeth and the well-known *Redriff* of Gulliver for Rotherhithe (*Colloquial English* 291). Curiously enough in the neighbouring county of Bucks, in Fingest (PN Bk 176), it is the initial *th* of *thing* which is changed to *f*. The change of initial *th* to *t* in this name may well have been due to official Anglo-Norman usage in connexion with the Hundred Court.

DAINTRY WOOD

In 1242–3 Walter FitzSimon held land in Tingrith (Fees 890). He seems to be the same as the Walter Fitz Simon who held land in Daventry or Daintry (Nth) (Fees 935), and if so this wood preserves a trace of his holding (cf. VCH iii. 436).

Toddington

TODDINGTON [tʌdiŋtən] 95 C/D 6

Dodintone 1086 DB
Totingedone 1086 DB
Tudingedon 1166, 1182 P, 1220 LS, 1232 FF, 1243 Cl
Tudingeden 1180 P
Tudingeton 1186, 1193 P, 1198 Abbr, 1229 Cl
Todingedon 1187 P, 1250 Ch, 1276 RH
Tudinton 1194 P, 1244 Cl
Tuddingeton 1198 Abbr, 1227 Ass
Tudingetun 1198 FF
Tundingedun (sic) c. 1200 (13th) *Dunst*
Totingedun c. 1200 (13th) *Dunst*
Tudington, Tudyngton 1219 FF, 1227 Ass, 1229 Ch, 1231 Bract, 1240 Ass, 1244 Cl, 1287 *Ass*, 1315, 1392 Ch

Tudingdon, Tudyngdon   1227 Ass, 1231, 1238 Cl, 1240 Ass, 1247, 1276 *Ass*, 1286 Dunst, 1287 *Ass*
Todingdon, Todyngdon   1227, 1240 Ass, 1276 *Ass*, 1291 NI, 1316 Ipm, FA, 1333 Ipm, 1390–2 CS
Todington, Todyngton   1227, 1240 Ass, 1276 *Ass*, 1316 Cl, 1489 Ipm, 1526 LS
Tuddingetun   1232 Cl
Tudendon   1232 Cl
Tudingesdon   1239 FF
Tutdingedon   1242 Fees 868
Tuddingdon   1244 Cl
Totingdon   1276 *Ass*
Totingedon   1276 RH
Totynton   1342 Ipm
Totyngdon   1382 Cl, 1439 IpmR
Tudyngdown   1388 Cl
Tuddington   1576 Saxton, 1766 J, 1780 Jury
Taddington   1662 Fuller

'Hill of Tuda's people' *v.* inga, dun. The forms with *o* were at first purely spelling variants but ultimately tended to affect the pronunciation except locally. Skeat is inclined to identify this with the place called *Tudincgatun* (Thorpe 527), but it is very improbable that that is in Beds and we should certainly expect *-dun* rather than *-tun* as the termination. Of considerable interest is the occurrence of the field-name *Tuddeworth* in this parish, in the Dunstable Cartulary. It is clearly named after the same man[1].

CHALTON

Chaltun   1131 (13th) *Dunst*
Chalton   1195 P *et passim*, 1287 Ass (*juxta Tudington*)
Chalftun   c. 1200 (13th) *Dunst* (p)

---

[1] With the aid of Mr Hight Blundell's *Toddington: its Annals and its People* we can trace the long survival of certain field-names found in the 13th cent. in the Dunstable Cartulary: *Cherswellemede, Hemoridie, Waperincge, Hickwelle, Levesethe* appear as *Carswell Meadow, Heymoriddy, Wapwinge, Hickwell, Lewsey* in the 1581 Survey of the Manor of Toddington. Similarly *Stonyham, Longspert, the Quabbe* of a 1453 deed appear as *Standingham, Longsperte* in the same Survey, while the last survive as Longspert and *Squabb* or *Quabb* near Cowbridge, to this day.

Chauton 1200, 1203 FF, 1227 Ass, 1236–7 FF, 1250 Ch
Chalfton 1227 Ass
Chaulton 1610 Speed, 1766 J

'Calf-farm' *v.* cealf, tun.

CONGER HILL

Cungar, Cunniger 1597
Cungar 1621

These early forms are quoted from Mr Hight Blundell's history of Toddington (p. 5) and point to ME *conynger*, 'rabbit-warren' (cf. NED s.v. *conyger*). There is no trace of such now, but it is perhaps hardly to be expected and the earthwork on the hill must have had some earlier name now lost.

EFFERIDDY (local)

Aelfredeswelredy 13th *Dunst*

This local survival is noted in BHRS v. 139. It is interesting as an example of the OE riðig, which is fairly common in the minor names of early Beds documents but which seems to have left no trace on the map. The whole name means 'stream from Alfred's spring.'

FANCOTT[1]

Fencote 1212, 1215 FF, 1276 *Ass* (p)
Fancote 1224 FF, 1276 *Ass* (p), 1304 AD

'Cotes in the marsh' *v.* fenn, cot.

FRENCHMANS WAY (local)

Franchemanewei 13th *Dunst*

This old name for the road through Tebworth to Hockliffe (Blundell, *Toddington*, 196) would seem to hide some historical incident lost beyond recovery.

HERNE (Grange, Green, etc.)

Hare 1183 P, 1202 Ass, 1220 LS, 1227 Ass, 1236 FF, 1238 Cl, 1242 Fees 868, 1247 FF, *Ass*, 1250 Ch, 1276 RH, *Ass*, 1302, 1316, 1346 FA, 1392 Ch, 1428 FA

---

[1] This is possibly the *East Coten* of BCS 659, cf. BHRS v. 168.

Haren 1276 *Ass*, 1286 Dunst, 1297 *SR* (p)
Todingdon Grange de Hare 1291 Tax
Harne 1535 VE
Hairn 1766 J, 1826 B

This name is difficult. It is clear that the modern form is corrupt and that the *Hairn* of the earlier maps is a good deal more correct. In considering the meaning of the name it should be noted *Hairn* probably applies to a largish area rather than to any single spot, for in Bryant's map it is printed right across the Toddington road with considerable spacing, and this agrees with the fact that we now have Herne Green, Herne Manor Farm and Herne Farm on the north side of the road and Herne Grange and another Herne Farm on the south side of the road. All alike are on high ground, Herne Manor Farm being at the highest spot in the district.

Professor Ekwall takes the name to be a dat. pl. and identical with Harome (Y), DB *Harum*. For this we must assume an OE word corresponding to Swed *har*, 'stony place,' *stenhar*, 'heap of stones,' MLG *hare*, 'height,' Dutch *haar*, 'height covered with wood' (often in p.n.s as *Haaren*). Cf. further Jellinghaus, *Die Westfälischen Ortsnamen*, s.n. *haar*. Hence, '(At the) *hars*,' but the exact sense of that word we cannot determine.

HIPSEY (Spinney) (6″)

Eppesho wood 1250 Ch, 1279 Misc.
Eppeho 1386 Pat
Ipsey bush 1607 *Terr*
Ipsow 1750 *Land Tax Assessment*

'Eopp's hoh' or 'spur of land.' The weak form *Eoppa* is well established as an OE pers. name and is clearly a pet-form of one of the OE pers. names in *Eorp-*.

WADLOW (lost)[1]

Wadelawe 1200 FF (p), 1202 Ass (p), 1212 FF (p), 1220 LS (p), 1236, 1243 FF, 1247 *Ass*
Waudelawe 1236 FF, 1240 Ass (p)

---

[1] Lysons says (143) that this was a mile from Toddington and that there were considerable traces of buildings in a field which goes by the same name. This is clearly the field known as *Wadlowes*, *Wadeloes*, in the 17th cent. strip map, near Redhill Farm.

Wadlowe 1236 FF
Wadelowe 1247 *Ass*, 1250 Ch, 1286 Dunst, 1291 Tax, 1323 Ch

Probably 'Wada's hill or barrow' *v.* hlaw. The two forms in *Waud-* look as if the first element were weald, but that would make the subsequent development very difficult to explain.

WARMARK

Warimarc, Waremerche 13th *Dunst*

There is not much to go upon here. The suffix is clearly OE mearc and the farm lies near the boundary of the parish. The first part may be an unrecorded OE pers. name *Wǽra*, a short form of one of the numerous OE names in *Wǽr-*. Alternatively it might be an OE *wearg(a)-mearc*, 'outlaw(s)-mark,' referring perhaps to a place where such were to be found or where bodies might be thrown after execution. *v.* mearc.

Totternhoe

TOTTERNHOE 95 F 5/6

Totenehou 1086 DB
Toternho(u) 1160–7 (13th) *Dunst*, 1202 Ass, 1220 (13th) *Dunst*, 1227 Ass, 1234 FF, 1247 *Ass*, 1284 FA, 1286 Dunst, 1304 Ipm, 1316 FA, 1323 Ch, 1333 Ipm, 1350 AD vi
Toterho(u) 1176 P, 1202 Ass, 1242 Fees 891, 1247 Cl, 1257 Ch, 1262 FF, 1302 FA
Thotherno, Tothirno 13th *Dunst*
Totenho 1227, 1240 Ass, 1247 *Ass*
Thoterho 1241 FF, 1286 Dunst
Toterno 1247 *Ass*, 1286 Dunst, 1287 *Ass*, 1299 Ipm, (*by Dunstaple*) 1327 Ipm, 1489, 1497 Ipm
Thoturno Hy 3 BM
Totorno 1332 AD i, 1352 BM
Toturnowe 1338 AD iii, 1347 Cl
Totourno 1393 BM
Toturnho 1412 AD vi, 1461 IpmR, 1463 AD i
Toturhoo 1428 FA
Totern(h)oo 1503 Ipm, 1526 LS
Tatternall 1576 Saxton

Totternhoe 1657 BM
Tatternhoe 1780, 1785, 1791, 1810, 1820 Jury

The name *Totternhoe* has long been considered a crux, at least so far as the first element is concerned. Consideration of the topography and archaeology of the site however provides the solution. The *hoh* is a promontory of the chalk downs, with very steep sides on the west and south but with a more gradual slope to the east. It is crowned by an ancient camp, commonly known as Totternhoe Castle and commands a view of Watling Street, which runs past it from south-east to north-west, just two miles away. All this suggests that the first element is OE *tot-ærn*, 'look-out house' (*v.* ærn). This compound is not on record in OE, but has its parallel in the familiar *Toot Hill*, *Tuthill*, so commonly used of a look-out hill. The castle was clearly used as a 'look-out house.' This interpretation is interestingly confirmed by Mr Goddard, writing in the VCH (i. 294) with no suspicion of the actual meaning of the place-name. He says 'the position is a majestic one, and to those moving on the lower plains for miles round, the Totternhoe mound seems to keep watch on its height like some great conning-tower[1].'

COOMBE (Fm) (6″)

Combes (p) 1276 *Ass* (p)

v. cumb. The farm lies in a depression on the south side of the hill.

DOOLITTLE MILL

v. Dolittle Mill *supra* 68.

EGLEMONT (lost)

Eglemunt 1166 P, 1171–7 (13th) *Dunst*
Eglemont 1318 Cl

The site of this quarry was in the fields north of Totternhoe (VCH iii. 448), and it was used to provide the stone for the

---

[1] Tattenhoe (PN Bk 73) has some forms in *Totern-* and, curiously enough, lies in the same relation to Watling Street some thirteen miles to the north-west, but the existence of early forms in *Tat-* forbids the same etymology and the forms in *Totern-* are probably to be explained as due to the influence of the name of the more important place.

king's house at Windsor (cf. Pipe Roll for 1166). The meaning is clearly 'eagle's hill,' a picturesque name given by the Normans to Totternhoe Hill.

LOWER END

Netherhende (p) 1287 *Ass*

Self-explanatory. In early days *nether* was used for such names rather than *lower* as now employed.

PASCOMBE PIT (6″)

Passecumbe 1264 (13th) *Dunst*

'Passa's cumb.' The combe is a hollow in the Dunstable Downs. The pers. name *Passa* is on independent record and is also found in Passenham (Nth). *v.* IPN 147 n. 4 for further notes on this pers. name.

Westoning

WESTONING 95 B 7

Westone 1086 DB *et passim*, (*Tregoz*) 1316 FA, (*Ynge*) 1365 Fine
Westonynge 1373 Cl
Westnyng, Westning 1518 *Award*, 1646 NQ iii

'West farm,' but why so called is not clear. The only place from which it is really west is Higham Gobion. This is a good four miles to the east. It is rather nearer to Barton-in-the-Clay, but this is rather south of west. The present form of the name is misleading, it should be Weston Ing, the Ing being manorial and due to the acquisition of the manor by Chief Justice *Ing* in the days of Edward ii (cf. VCH iii. 452). The Tregoz family held the manor for a time in the 13th cent. (ib.).

CLAYHILL

Clayhull 1460 AD iii

'The soil (of Westoning) is varied, on a subsoil of strong clay' (VCH iii. 451).

SAMSHILL

Samsill 1617 BHRS ii. 108
Samsell 1660, Bunyan, *Relation of Imprisonment*, 1766 J

Whipsnade

WHIPSNADE 95 G 6

Wilbesnede 1202 Ass
Wibsnede, Wyb- 1202 FF, 1287 *Ass*
Wybbesnathe 1227 FF, 1332 AD i
Wibbesnethe 1227 FF
Wyppesnade 1227 Ass, 1308 AD i, 1382 Cl
Wylbesnethe 1227 Ass
Wibbesnade, Wybb- 1236 FF, 1242 Fees 869, 1247 *Ass*, 1252
   FF, 1291 NI, 1319 AD vi, 1333 AD vi
Wybesnade, Wib- 1236 FF, 1291 NI, 1316 FA, 1340, 1349,
   1350 AD vi
Wipesnade, Wylbesnad, Wybesnathe 1247 *Ass*
Wybbesned 1247, 1276 *Ass*
Wybbesnet 1276 *Ass*
Websnade 1299 Ipm, 1374 AD vi
Wypsnade, Wip- 1352 BM, 1390–2 CS, 1394, 1490 AD vi,
   1491 Ipm
Websenad 1374 AD vi
Wipsonade 1421, 1422 AD i, 1441 AD ii
Wyppysnade 1526 LS

'Wibba's snæd,' the reference being probably to the piece of
woodland cut off and cleared by the original settler. It is very
doubtful if this name is found on independent record in OE,
the *Wibba* of Florence of Worcester being an error of transcription for *Pibba*, but it is clearly found in *Wibbandun* (ASC
s.a. 568) and must be a pet-form for some such name as *Wīgbeald*
or *Wīdbeald*. There was also a *Wybbesethe* in Whipsnade in
1308 (AD i) which is a compound of seað with the same pers.
name.

DEDMANSEY (Wood) (6″)

Dudewineshei 1227 ClR
Dudeuinishei 1254 BHRS viii. 211
Dodesynnishay 1299 Ipm
Dodenynishey 1333 AD vi
Dudmansheye Eliz ChancP
Dudmanse 17th VCH iii. 455

'Dudewine's *hay* or enclosure' *v.* (ge)hæg. This name, as also Shortgrove, reminds us that we are in wooded country and confirms the suggestion made above as to the exact significance of the *-snade* of Whipsnade.

THE GREEN

Wybesnadegrene 1350 AD vi
le comune grene 1409 AD i

Self-explanatory. 'The village is grouped round a large green' (VCH iii. 455).

SHORTGROVE

Sortegraue 1160–7 (13th) *Dunst,* 1209 FF
Scortegrave c. 1160 (13th) *Dunst*
Shortegrave 1209 Abbr
Schertegrave 1304 AD ii, 1305 AD iii

Self-explanatory.

Woburn

WOBURN [wu·bən] 95 B 4

Woburninga gemære 969 BCS 1229
Woberne 1086 DB
Woburne 1086 DB *et passim*
Wobburn 1130, 1159, 1164 P
Weburna 1170 P
Wuburn 1202 Ass, Hy 3 BM, 1227 Ass, 1235 Cl, 1240, 1241 FF, 1245 Ch, 1247 *Ass,* 1252 FF, 1268 Pat
Wouburn 1202 Ass, 1271 FF, 1295 Ipm, 1299 Ch, 1305, 1320, 1334 Cl, 1337 Ipm, Cl, 1364, 1381 Cl, 1389 IpmR
Wburne 1227 Ass
Wubburn 1236 FF
Woughburne 1307 *Ass*
Wobourne 1332, 1364, 1376–7 Cl
Woubourn 1332 Ipm, Cl, 1334, 1381 Cl
Wooburn 1526 LS
Woburn or *Woobourne* c. 1750 Bowen

'Winding stream,' a compound of OE **woh,** 'twisted,' and **burna.** Cf. Woburn (Sr, W).

BIRCHMORE

> *Birchemore* 1227 Ass, 1242 Fees 867 *et passim*
> *Birchmore* 1299 Ipm *et passim*

HORSE POND (6″)

> *Horspol* 1276 *Ass*

These two are self-explanatory.

LOWE'S WOOD

> *Losewod* a. 1340 VCH iii. 459
> *Louze Wood* 1826 B

It seems clear that the modern form is corrupt and that the name really goes back to OE *hlos-wudu* (BCS 627) which it is difficult to think can be anything else than 'pigstye-wood,' *v.* hlose. There was another *Loswude* in Chicksand (*Warden* 19 *b*).

UTCOATE GRANGE[1]

> *Utcote* 1276 *Ass* (p)
> *Uttecote* 1299 Ch, 1306 Abbr
> *Vutecote* 1331 QW

'Out-cottages' *v.* cot. The grange lies in the south-west corner of Woburn parish and is probably so called from its remote position.

WHITNOE (Orchard) (6″)

> *Wytenho, Witenho* 1291 Tax, 1307 *Ass*, 1337 Pat
> *Whitenho* 1299 Ch, 1331 QW

OE *hwītan hō* (dat.), 'white spur of land,' *v.* hwit, hoh.

VIII. FLITT HUNDRED

> *Flictham* 1086 DB
> *Flicte* 1179 P
> *Flete* 1185 Rot Dom
> *Flitte* 1193 P, 1202, 1240 Ass, 1247 *Ass*, 1284, 1316 FA, 1327 Ipm, Fine, 1328 Fine, 1390–2 CS, 1428 FA

---

[1] The new Popular 1 in. O.S. map has *Ulcoate*, but this must be a mistake. The 6 in. map has the above form, which agrees with the earlier history of the name.

Flitton 1297, 1327 Ipm, Orig
Flitten(e) 1297 *SR*, 1327 Fine, 1329 Pat, 1332 Ipm
Fletton 1297 Cl
Flytte 1302, 1346 FA, 1495 Ipm
Flett 1310 Misc, 1587 D
Flittescourt 1316 FA
Flite 1327 Orig
Flute 1353 Orig
Flutt 1408 IpmR

For a discussion of the etymology of this name, *v.* Flitton *infra* 148. The site of the meeting-place of the Hundred is unknown. It must have been in the neighbourhood of Flitton or along the streams which gave rise to that name. This, as in Manshead Hundred, makes the meeting-place to have been roughly at the middle of its longer axis, very near to the meeting-place of that Hundred itself. The 'Thingrithe' of Tingrith (*v. supra* 134) is a western extension of the Flitt streams.

Barton-in-the-Clay

BARTON-IN-THE-CLAY 95 C 9

Bertone 1086 DB, 1189 P
Barton 1202 Ass (p) *et passim*
Barton iuxta Shuttelyngton c. 1370 *Linc*
Barton-in-the-Clay 1535 VE
Barton-le-Clay c. 1560 *Linc*
Barton Clay 1675 Ogilby

v. bere-tun. The soil is a strong clay, with a subsoil of clay and chalk (VCH ii. 308).

BROOKEND GREEN

cf. *le Brokfurlong* 1287 *Ass*

Self-explanatory.

Caddington

CADDINGTON 95 F 8

Cadandun c. 1053 (c. 1250) KCD 920
Cadendone, Kadendon 1086 DB, 1131–3 (13th) *Dunst*, 1181, 1222 St P, 1247 *Ass*, 1286 Dunst, 1287 *Ass*, 1316 FA, 1345 AD vi, 1350 AD i
Cadendun 1189 (1227) Ch, 12th St P, 1222 BM

Cadingdun 1195 FF
Cadinton 1202 'Ass (p), 1276 *Ass*
Cadyndon, Kadindon 1227 Ass, 1242 Fees 867, 1299 AD iii
Cadingdon, Kad- 1247 *Ass*, 1286 Dunst, 1390–2 CS, 1394 AD vi
Cadington, -yng- 1247, 1287 *Ass*, 1353, 1367, 1382 Cl, 1450
   IpmR
Westcadyngton 1353 Cl
Carington 1563 *Deed* (*pen*. Dr G. H. Fowler)
Carrington 1690 NQ i

'Cada's hill' *v*. dun. For the old local pronunciation, cf.
Doddington (Nb) pronounced as *Dorrington* (PN NbDu 65) and
Derrington (St) and Derrythorpe (L) which both originally
had medial *d*. We have the reverse process in Paddock Wood
(K) from *Parrok* and dialectal *poddish* for *porridge*. The name
occurs again in this county in Cadbury *supra* 56.

SKIMPOT (Fm)

This name is said to be a corruption of 'St Mary's Pottery'
(*v*. W. G. Smith, *Dunstable*, 131), in the same way that *Skimmery
Hall* was colloquial for St Mary's Hall in Oxford.

ZOUCHE'S FARM

This seems to represent a manor held of the dean and
chapter of St Paul's by the family of Zouche of Harringworth
(cf. VCH ii. 315).

Clophill

CLOPHILL 95 A 9

Clopelle 1086 DB
Clophull(e) 1227, 1240 Ass, 1242 Fees 889, 1250–1 FF, 1273
   Cl, 1276, 1287 *Ass*, 1291 NI, 1293 Ch, 1297 *SR*, 1302 FA,
   1307 *Ass*, 1316, 1346 FA, 1350 Ipm, 1354 BM, 1359 Cl,
   1360 Ipm, 1384 Cl, 1390–2 CS, 1428 FA
Clop(p)hill 1247, 1287 *Ass*, 1361 Cl
Claphull 1287 *Ass*
Cloppehull 1287 *Ass*
Clouphull 1287 *Ass*, 1362 Cl
Clapwell 1654 Cai
Claphill al. *Clapwell* 17th BHRS viii. 146

'Tree-stump hill.' For the element *Clop-, v*. Clapham *supra* 22.

BEADLOW

Beaulieu 1254 FF, 1293 Ch, 1310, 1393 Cl, 1529 LP
Bello Loco 1254 FF, 1287 QW, 1390–2 CS
Bealliu c. 1270 Gerv
Beleu 1428 Chron St Alb
Bedlowe, Beawlowe 1578 BM
Bedlow 1590 BM

An interesting anglicising of an earlier French name. The nearest parallel is the development of the same French name to Bewdley (Wo). OFr names in *Beu-* developed in later English either to *Bew-* or *Bea-*. Here, as in the Bucks *Beamond* End, we have the second of these two developments. The appearance of a name of French origin in this parish must be connected with the foundation of this cell of St Alban's by the family of Albini who early had a castle at Cainhoe.

CAINHOE

Chainehou, Cainou 1086 DB
Cahenno c. 1115 (c. 1200) Hist. Mon. de Abingdon ii. 101
Kainho, Cainho, Caynho 1227 Ass *et passim*
Keinho, Keynho 1235 Fees 460 (p), 1287 Cl, Ipm, 1311 Ipm
Cayno, Kayno E 1 KF, 1272 Ipm, 1275 Cl, 1276 *Ass* (p) 1346 FA, 1380 Cl, 1428 FA
Kayho, Cayho 1272 Ipm, 1287 *Ass*
Keynoo 1512 AD vi
Keynho Eliz ChancP

' Cæga's hoh ' or ' spur of land.' This is the weak form of the pers. name *Cæg* found in Keysoe *supra* 14. On *Cæg* and *Cæga, v.* IPN 180.

MODDRY or MODDRI (lost)

This suspicious-looking name is given in Dugdale iii. 276 as the original name of Beaulieu Priory in the abstract in the St Albans Register of a document datable as before 1146. In this it occurs several times: the other place-names in the document are fairly good, except *Rinethella* for *Amethella*.

PEDLEY WOOD

Pudelegh 1324 Ipm

'The leah or clearing of *Pydda*.' Without the intermediate forms it is difficult to be sure of the history. Cf. Piddington (Nth) and Pidley (Hu) *infra* 211.

Flitton

FLITTON 95 A 8

Flichtham 1086 DB

Flitte, Flytte 1166 P, 1202, 1227 Ass, 1236 Cl, 1242 Fees 891, 1270 FF, 1276 *Ass*, 1282 Ipm, 1284 FA, 1287 *Ass*, 1302 FA, 1305 Ipm, Cl, 1316 Ipm, FA, 1329 Ipm, Cl, 1388 Cl, 1390–2 CS, 1396 Cl, 1428 FA

Flete 1183 P

Fliuten[1] c. 1200–25 BHRS ii. 128

Flitten c. 1200–25 BHRS ii. 128, 1276 *Ass* (p), 1279 RH (p), 1297 *SR*, 1327 Ipm

Flette 1240 Ass

Flyten 1260 FF

Flyte 1262 FF

Flit 1286 Dunst

Flute 1291 NI

Flitton, Flytton 1318 Ch, 1331 QW, 1526 LS

Flyten 1331 QW[2]

It is clear that in the discussion of this difficult name we must take into account, not only the forms of Flitton itself but those of Flitt Hundred, which must be associated with it, and of Flitwick just across the valley from Flitton, in the Hundred of Redbornstoke. Further, one must note that the stream which divides Flitton and Flitwick parishes is known as the River Flit(t), though the only actual record of this on the O.S. map is its expansion to a lake known as *Flit* Water, in the grounds of Flitwick Manor. Also, there still survives the Rural Deanery of *Fleete*, which is recorded in 1291 as *Flitte, Flute* (NI) and *Flett* in 1535 (VE).

---

[1] Printed as *Flinten*, but this must be an error for *Fliuten* or, possibly, *Fluiten*.

[2] The forms in the *Lincoln Registers* are *Flitt(e), Flytte* (1290–1431, six times), *Flette* (c. 1440, twice), *Fleytt* (c. 1440), *Flytt(e)* (c. 1470, four times), *Flitton* (c. 1480).

Topographically it is to be noted that there are in reality a number of small streams parallel to one another, with small connecting streams, rather than a single stream, running along the valley which separates first Pulloxhill and then Flitton from Flitwick.

It is clear that the names in question must have something to do with OE *flēot* or *flēote*, used of a small stream, the most familiar example of which is the one which gave its name to Fleet Street in London. Such would explain forms like *Flete*, *Flieteuuiche*, *Fletwyk*, *Flettewyke*, *Flotewyk*, *Flutewyke*, *Fliuten*, *Flute*, for we might expect such developments from OE *ēo* in ME, the variant forms representing variant dialectal developments of the same OE word. They would clearly also account for the modern Deanery name. We are still, however, left with the puzzle of all the early forms in *Flit-* and still more with those in *Flitt-*.

Professor Ekwall suggests for the solution of this puzzle that there may have been very early shortening of ME *Flīt* from OE *Flēot* (cf. *Liefric*, *Liueua* in 12th cent. Beds Fines from OE *Lēofric*, *Lēofgifu*) to *Flĭtt*, helped by the influence of Flitwick, in which such shortening would be natural. The *c* of DB he takes to be a mistranscription for *e*. Traces of this form may also be found in the form (*on*) *flitum* found as a gloss for Latin *fluctris*. This, Toller suggests (B.T. Supplt. s.v. *fleot*), is a mistake for *flactris*, *flactra* being itself a Latin word for *locus coenosus*. This sense of 'marsh' would perhaps suit the ground even better than stream[1] but, however that may be, the dat. pl. form *flitum* serves to remind us of the origin of the curious *Fliuten*, *Flit(t)en(e)* forms which lie behind the modern Flitton. The usual form of the place-name in the dat. sg. would be *flēote* in OE, but as there was more than one stream or marsh, equally common would be the dat. pl. form *fleotum* and these in ME would develop to forms with a suffix *-en*[2]. The rare suffix *-en* was naturally altered to *-on* in later days. We are left with the small problem of the DB form for *Flitton*. The *-ham* may be simply for *-am*, the Lat. fem. accusative suffix which

---

[1] In the *Annals of Dunstable* (108) we have a reference to *Mariscus* de Flitte.

[2] For such double forms cf. Ion *infra* 151.

we sometimes find added to place-names in DB when they are in the accusative after 'tenet' (cf. *Evreham* in DB for Iver (Bk)). Alternatively the *-ham* may be a bad form for the dat. pl. suffix *-um*. These names therefore mean 'wic by the stream,' and '(at) the stream or streams.'

GREENFIELD

> *Grenefelde* 1286 Dunst
> *Grenfelde* 1548 Pat
> *Grenfield* 1766 J

Self-explanatory.

WARDHEDGES

> *Wardegges* 1276 *Ass*
> *Wardhegges* 1276 *Ass* (p), 1297 *SR* (p)

'Protecting hedges' v. **weard**. The second element is OE *hecg*, 'hedge.' The same name is found in Weybridge (Hu) as *Wardehegges* (Ramsey Cartulary, ii. 301). In both places it probably has reference to some game-enclosure.

WORTHY END

> *Wrthing* 13th *Dunst*, 1240 Ass
> *Worthinge* 1276 *Ass* (p), 1297 *SR* (p)
> *Westworthinge* 1286 Dunst, 1302 FA
> *Worthinge de Flitte* 1287 *Ass* (p)

This p.n. may have the same history as that suggested by Ekwall for Worthing (Sx) and be from OE *Weorðingas*, 'Weorð's people,' a derivative of a lost OE pers. name (PN in *-ing* 64) which he finds also in Worston, Worsthorne and Worthington (La). Alternatively, and perhaps more probably, this place and the Sussex Worthing may be from OE *worðingas* and mean simply 'the people of the worð or enclosure.' For such names cf. the citation under Eastcotts *supra* 91.

Gravenhurst

GRAVENHURST 95 A 9/10

> *Crauenhest* 1086 DB
> *Gravenherst* 1206 FF (p), 1504 Ipm
> *Graveherst* 1213 FF, 1227 Ass

Gravenhurst 1223 FF *et passim*
Gravehurst 1225 Pat, 1302 FA
Grauenhirst 1227 Ass
Grauehirst 1227 Ass, 1260 FF
Grauenhirste, -hyrst 1232 FF, 1266 AD i (*Parva*)
Grafhurst 1240 Ass
Great Gravenhyrst c. 1270 AD i
Grauynherst 1287 *Ass*
Cravenhurst 1377 BM
Grovenhurst 1428 FA
Overgravenest 1534 BM

Professor Zachrisson suggests OE *grāfan-hyrst*, 'wooded hill of the grove or thicket[1],' with the rare *grāfa*, 'grove,' found in *ellen-grāfa*, 'elder-grove' (B.T. *s.n.*). The *a* would be shortened in the trisyllable, with sporadic preservation of the long vowel and development to *grove*. The vowel is now long but that may be due to ready association with *grave*.

ION

Eie, Eye 1202 Ass (p) *et passim* to 1413 IpmR
Eyeyn 1260 FF
Eyen 1270 FF (p), 1287 *Ass*, 1297 *SR*, 1323 Ipm, 1428 FA (p)
Eyon 1492 Ipm, 1637 NQ iii
Yon 1504 Ipm
Yen 1549 Pat

'(At the) island' or '(at the) islands,' the two forms of the name being the dat. sg. and pl. of eg. The sites must have been 'islands' in the marshes which once covered this low-lying and well-watered ground[2]. For such variant forms, cf. Flitton *supra* 148.

Haynes

HAYNES [heinz], [hɔ·nz] 84 J 9

Hagenes 1086 DB, c. 1206 BHRS i. 104, Hy 3 (1317) Ch
Hagnes c. 1150 BM

[1] The Rector, the Rev. H. J. Baylis, writes 'obviously the parish was once thickly wooded in the higher ground and it consists mostly of a hill with its slopes.'
[2] Great Ion Farm is still subject to floodings after heavy and continued rain (*ex inf.* the Rev. H. J. Baylis).

Hawenes 1202 Ass (p), 1219 FF, 1291 NI, 1297 *SR*, 1302, 1316 FA, 1347 Pat, 1390–2 CS
Hawenis 1219 FF, 1276 *Ass*
Haunnes 1242 Fees 887
Hawnes 1247 *Ass*, 1287 *Ass* (p), 1426 IpmR, 1535 VE, 1549 Pat, 1800, 1830 Jury
Haunes 1247 *Ass*, 1257 FF, 1276, 1287 *Ass*, 1291 NI, 1326 Cl, 1327, 1351 Ipm, 1361 Cl, 1402 CS, 1610 Speed
Hauwnes 1287 *Ass*
Hawens 1325 Cl, 1780 Jury
Haynes c. 1560 *Linc*
Hains 1712 NQ i

This is a difficult name and one can only offer tentative solutions of it. There is, quite apart from the mention of the hero *Hagena* in *Widsith*, adequate evidence for an OE pers. name *Hagena*, and Björkman (ZAN 42) seems right in accepting it as against Redin (97) and Forssner (139) who are sceptical about its being anything but continental. With this pers. name as a basis one can perhaps take this name as consisting of *Hagenan* (gen. sg.), followed by næss, applied here to a spur of land. It is not unsuited to the site of Haynes, the nucleus of which we may assume to have lain along the ridge on which the church stands. For the use of this word for an inland site, cf. Nazing (Ess) in PN in -*ing* 47. *Hagonannæss* would inevitably become *Hagenes, Hawnes* in ME.

One other possibility may however be suggested. The OE **haga**, 'hedge,' was in OE and, in the form *hawe*, in ME, used of an enclosure, a yard, a messuage. The original name may have been *haga-næss*, 'haw-ness,' and have been descriptive of a spur of land on which stood a 'haw.' There is a curiously apposite parallel for such a name in the Swedish Haganäs, dealt with in *Sverges Ortnamn* vii. 1. 236. The modern spelling and pronunciation are entirely arbitrary, though perhaps influenced by the alternative *hay* = hedge.

NORTHWOOD END

Norwd' 1202 Ass
Northwode 1204 FF (p)
Self-explanatory.

Higham Gobion

HIGHAM GOBION [gaubiən] olim [gʌbin] 95 B 9

Echam 1086 DB
Heham 1166 P
Hecham 1231 Cl
Hegham 1287 Ass
Heygham Gobyon 1291 NI
Higham Gubyns c. 1520 Linc
Higham Gubion 1526 LS
Higham Gubbin(s) 1595 Cai, 17th NQ i

'High homestead,' aptly descriptive of its site, v. heah, ham.
The Gobion family were associated with the manor from the
12th cent. on (cf. VCH ii. 345). For the popular development
of this family-name, cf. the history of Gubeon (PN NbDu 97).

FALDO

Faldhou c. 1125 (c. 1350) Rams (p)
Faldehou 1166 P
Faldho 1168 P, 1227, 1240 Ass, 1247 Ass
Faudho 13th Dunst, 1227 Ass, 1247 Ass
Faudeho 1202, Ass (p), 1227 Ass
Faldehow 1227 Ass
Faudo 1247 Ass
Faldo 1239 FF (p), 1297 SR
Faltho 1336 Fine, 1338 Cl

'The hoh or promontory of land with a fald or fold upon it.'

Hyde

HYDE (EAST and WEST) 95 G 10

la Hide 1197 FF (p)
Esthide 1247 Ass
Westhide 1276 Ass
Esthith, Westhith 1287 Ass
v. hid.

DANE END (Fm)

de la Dane 1297 SR (p)

'Valley end' v. denu. ME dane for dene is occasionally found,
coming from an OE form dænu in place of the more usual denu.

FERNELS WOOD (6")[1]

Fenelsgrove or *Fennel Grove* al. *Fenels Luton* 1504 Ipm

This goes back to the FitzNeel family who, as early as 1283, held land in Luton (VCH ii. 358), and this wood formed part of their manor of Fennels Grove.

KIDNEY WOOD

Cutheno 1198 FF
Cutenho, Kutenho 1199 FF, 1307 *Ass* (p)
Ketenho 1201 FF
Cothernhoe 1772 BHRS v. 102
Kitnoe 1794–1841 ib.
Kidnoe 1811 ib.
Kidney 1831 ib.

These forms point to an OE *cytan-hō*, 'kite's hoh,' the hill being so called because frequented by a kite. It is however possible that the first element is a pers. name *Cyta*, identical with the name of the bird.

THRALESEND

' In 1390 Nicholas de la Haye confirmed to his mother, Agnes Thrale, lands in West Hide, Luton' (VCH ii. 359).

Leagrave

LEAGRAVE 95 E 8

Littegraue 1224 ClR, 1227 Ass, 1276 *Ass* (p)
Littlegraue 1227 Ass, 1287 *Ass*
Lihtegraue 1227 Ass
Leytegraue c. 1250 *Deed* (*pen.* Dr G. H. Fowler)
Lithtegraue 1276 *Ass*, 1286 Dunst
Lithegrave, Lyth- 1276 RH (p), 1287 *Ass*
Lytegraue 1286 Dunst, 1379 Cl, 1390–2 CS (p)
Lyghegraue 1287 *Ass*
Lytelgraue 1287 *Ass*, 1379 Cl
Lightgrave 14th Gest St Alb, 1461 IpmR, 1499 Ipm
Lygrave 1504 Ipm, 1610 Speed, 1675 Ogilby

---

[1] The form on the modern O.S. map seems to be an error.

Lighgrave 1504 Ipm
Lyttgrave 1535 VE

It is at first sight difficult to bring the different forms of this name into coherent relationship. In particular, the forms which suggest that the first element is OE *lytel*, 'little,' cannot at first be reconciled with those which point to OE *liht, leoht*, 'light.' But the difficulty disappears if it may be assumed that the first element consists, not of a significant word, but of a pers. name *Lihtla*, derived from a stem *Liht-*, which is known to have been employed, though rarely, in OE. The compound *Lihtweald* occurs in the LVD, and *Lihtwine*, in the form *Lictuin*, is found in the 11th and 12th centuries. An original form *Lihtlan grāf* will account for all the forms earlier than the 16th cent. *Little-* in 1227 can be derived from *Lihtle-* by assimilation of *ht* to *tt* just as OE *Witta* is best regarded as an assimilated form of *Wihta*. The spelling *Leyte-* can safely be taken as representing *Leyta-*, with *yt* for *ht*, from the parallel form *Leohte-*. The loss of *l* after *t* has many parallels. The influence of the river-name *Lea*, which first becomes apparent in 1504, did not finally prevail until recent times. *v.* **graf.**

LEWSEY

Leveseye 1291 Tax
Leveslhey 1535 VE
Levssey 1549 Pat

OE *Lēofes-ēg*, 'Leof's island,' the site being a well-watered one.

Limbury

LIMBURY 95 E 8

Lygeanburg 571 A (c. 899) ASC
Lymbiri, Lim- c. 1225 (13th) *Dunst*, 1276 *Ass*
Limbury, Lym- 1227 Ass, 1247 *Ass*, 1286 Dunst *et passim*
Lumbur(y) c. 1250 *Deed* (*pen.* Dr G. H. Fowler), 1276 *Ass*,
    1290 AD iv, 1307 *Ass* (p)
Lunboreye 1252 Ch (p)
Lumbyr' 1276 *Ass* (p)
Limberge 1296 Cl
Lymbery 1307 *Ass et passim*

Limbrey 1780 Jury
Limbury come Biscott (sic) 1785 Jury

'The fort of the river Lea' *v.* Lea, R. *supra* 8. This, and not the traditional Lenborough (Bk), is undoubtedly the correct identification of the place mentioned in the ASC as cited above.

BISCOT

Bissopescote 1086 DB, 1199 (1301) Ch
Bishopscot 1227 Ass
Bis(e)cot 1227 Ass, 1247 *Ass*, 1249 FF, 1276 *Ass et passim*
Bischecote 1247 *Ass*
Bishopescote 1287 *Ass*
Byssekote 1287 *Ass*
Bisshcote 1401 IpmR

'Bishop's cottages' *v.* cot. Who the bishop was is quite unknown, nor indeed need the word contain *bishop* in the official meaning of the word. There are two clear examples of *Biscop* as an OE pers. name, one of them in the genealogy of the kings of Lindsey, the other in the name of *Biscop Baducing*, commonly called *Benedict Biscop*. The latter was born at the very time of the conversion of Northumbria, and the early kings of Lindsey seem to belong to the same generation. Such a name, once established, might well survive for a long time without definite consciousness of its significance.

Luton

LUTON 95 F 9

Lygetun 792 (c. 1250) BCS 264
Lygtun 917 A (c. 925) ASC
Ligtun 914 D (c. 1050) ASC
Loitone 1086 DB, 1158 P
Luitun, Luytun 1156 P, 1279 Ipm, c. 1300 Gaimar
Luiton(a), Luyton 1161 P *et passim* to 1415 BM
Luton 1195 FF *et passim*
Lutton 1240 Ass, 1342 Ipm, 1381 Cl
Louytone 1276 *Ass*
Lytone E 1 BM
Luton al. *Lutton* 1288 Ipm
Lowton 1291 Tax

Leuton 1293 Cl
Luton al. *Luyton* 1304 Ipm
L(o)ughton 1376 Cl
Luton Soke Hy 5 AD iii
Luton Soken 1590 BM

'Farm by the Lea river' *v.* Lea, R. *supra* 8. For the terms
Soke and *Soken, v.* Eaton Socon *supra* 54.

BRACHE (Fm) (6")

la Brache 1276, 1287 *Ass*, 1422 IpmR
la Breche 1240 Ass, 1247 *Ass*, 1426 IpmR
le Breyche 1376 IpmR

v. bræc. For this term, cf. Breach in Maulden *supra* 81.

BRAMINGHAM (GREAT and LITTLE)

Bramblehangre 1240 Ass (p)
Brembelhanger 1247 FF, *Ass*
Bremelhangre 1247 *Ass*, 1269, 1271 FF, 1287 *Ass*
Brambelhangre 1247 *Ass*
Bramhangre 1247 *Ass*
Brumelhangre 1276 *Ass*
Brimelhangre, Brymelhangre 1276 *Ass*
Bramleshangre 1276 *Ass* (p)
Bramelhangre 1287 *Ass* (p), 1324 Ch, 1504 Ipm
Bremerhangre 1287 *Ass* (p)
Brambelhanger 1290 AD iv, 1297 *SR* (p)
Brymbelhanger 1379 Cl
Brambleangre 1426 IpmR
Bramanger 15th HMC, Var iv
Braminger 1599 NQ i

It is stated in the VCH (ii. 363) that these two forms represent
the ancient manor of Bramblehanger. If that is the case, it is
clear that the old name must have undergone some archaising
process, for the form *Bramingham* has not been traced back
earlier than 1570 (cf. BHRS ii. 108). The old manorial name is
a compound of OE bræm(b)el and hangra and the whole name
means 'wood on a slope, with brambles growing in it.' The
original site was perhaps the wood called now 'Great Bramingham Wood.' The normal development of this OE name is

found a few miles away in *Bramagar* Wood in East Hyde. The development of that name would however seem to be entirely independent of the manor in question.

COWRIDGE END [skəˑdʒend]

> *Curegge* 1196 Whet i. 421
> *Kuruge* 1202 FF
> *Curruge* 1247 *Ass* (p)
> *Courigge* 1276 *Ass*, 1327 Ipm
> *Courugg(e)* 1297 *SR* (p), Ipm
> 'Cow-ridge' *v*. cu, hrycg.

DALLOW (Fm) (6″)

> *Dolhou* 1247 *Ass* (p)
> (*la*) *Dolowe* 1287, 1307 *Ass* (p)
> *Dalleye* 1359 Cl

The first element in this p.n. is possibly dal and the whole name descriptive of a hoh or spur of land which was held in common but in which various persons held portions or *doles*. Alternative forms in which OE *dāl* has become *dole*, the long vowel being kept under the influence of the independent word, and in which the OE long *a* was shortened in the compound, before it became *o*, account for the variant developments.

THE DOWNS (6″)

> *la Dune* 1276 *Ass* (p)
> *la Doune* 1297 *SR* (p)
> Self-explanatory.

FARLEY (Fm and Hill)

> *Ferleya juxta Lectonam* Hy 2 (1285) Ch
> *Farlege* 1201 FF, 1297 *SR* (p)
> *Fareleye* 1276 RH, 1526 LS, 1590 NQ iii
> *Farle* 1291 Tax
> *Fayreley* Eliz ChancP
> *Farl(e)y* Eliz ChancP

Though no *n* is found in the early forms this is clearly for OE fearnleah, 'fern or bracken-covered clearing.' Farley (Db, Sf, Sr, W, Wo) and Farlow (Sa) all have *fearn* as their first element, and in all alike forms without an *n* are found from early times.

GREATHAMPSTEAD (lost)

   Grehamsted 1202 Ass (p)
   Grethamstede 1240 Ass, E 1 AD i, 1309 Ipm
   Grethamstude 1276 *Ass* (p)
   Gretehamstede 1504 Ipm

From OE great with second element hamstede, 'great *ham-stede*,' or from OE greot from the soil.

LUTON HOO

   le Hoo 1276 *Ass*
   Hoo in Luton 1480 Ipm
   Luton How 1782 Coll

v. hoh. Luton Hoo stands on a hill running down toward the Lea.

RUNLEY (Woods) (6″)

   Rindele, Ryndele 1202 Ass (p), 1276, 1287 *Ass*, 14th Gest
     St Alb
   Reyndel 1287 *Ass*
   Rundele 1287 *Ass*

No solution can be offered of this name. From the point of view of form the first element might be OE *rynel*, 'runnel,' but there seems to be no stream by the wood. (In that case we should have to take the *d* as epenthetic.) We have in OE a *hrindan broc* BCS 466 and *hrinde bearwas* (*Beowulf* 1363), now generally taken to be for *hrindede bearwas*, i.e. 'frost-covered groves.' Neither of these words would however explain ME *Rundele* or the modern *Runley*. These and the other forms point to initial *Hry-* or *Ry-* in OE. Professor Ekwall suggests an OE *rӯmde*, 'cleared,' hence *rӯmdan leage*, 'cleared leah.'

STAPLEFORD (lost)

   Stapelford 1274 Cl
   Staplefordefolde 1341 BM
   Self-explanatory.

STOCKWOOD (olim *Whyperley*)

   Wypereleya Hy 2 (1285) Ch
   Wyperley 1156 Dugd vi. 11, 1393 IpmR

Whip(er)ly Eliz ChancP
Stockwood al. *Wyperley* 1640 BHRS v. 110

Professor Zachrisson suggests that the first element in this name may be the *whippultre* (v.l. *whip(p)il-, wypul-*) in Chaucer's list of trees in the *Knight's Tale* (2065). That is usually taken to be the cornel-tree. For such a reduction we may compare Apperley (Nb) from *æppeltreo* and possibly Mapperley (Db) from *mapeltreo*. This would fit well with the suffix **leah.**

WOODCROFT (lost)

Odecroft[1] 1086 DB
Wodecroft 1297 *SR*, 1372 IpmR
Self-explanatory.

Pulloxhill

PULLOXHILL

Polochessele 1086 DB
Pollokeshill 1205 FF, 1287 *Ass*
Pollukeshull 1214 Abbr
Pulocheshella Hy 3 (1315) Ch (p)
Bulluckeshulle 1227 Ass
Pollokeshull 1228, 1236 FF, 1276, 1287 *Ass*, 1297 *SR*, 1337 Ipm
Pullokeshull 1242 Fees 890, 1260, 1270 FF, 1286 Dunst, 1313 BM, 1323 Ch, 1337 Ipm, 1346 FA, 1390–2 CS
Bolokeshill 1276 *Ass*
Pullukeshull 1286 Dunst
Poleshill 1287 *Ass*
Pullokeshill 1287 *Ass*
Pollekeshull 1337 Ipm
Pollexhulle, Polloxhulle 1428 FA
Polluxhull 1526 LS
Poloxhill 1577 AD iii
Pullockshill Eliz ChancP

One cannot get further with this name than Skeat's suggestion that the first element is a lost pers. name *Pulloc*, closely allied to ModE *Pollock*. The *Pollok-* forms may indeed contain that name or it may be that they are originally Anglo-Norman spellings with *o* for *u*, which for a time affected the pronunciation. Hence, 'Pulloc's hill,' *v.* hyll. Cf. Poulston (D), DB *Polochestona*.

---

[1] In DB it is an old hundred name and the identification is not certain.

KITCHEN END

> Kechyng, Kechinge 13th Dunst, E 1 BM
> Kechingges 1286 Dunst
> Kechyngg 1313 BM, 1331 QW

The only parallel to this difficult name which has been noted is Kitchingham (Sx) with a form Kechenham in 1242 BM. Professor Zachrisson suggests the possibility of an OE *Cyccingas containing the pers. name *Cuca found in Cucan healas (BCS 936), with e for i owing to the influence of the palatal ch.

UPBURY

> Hubberia 1166 P
> Hutberia 1168 P
> Hutteberia 1174 P
> Upbiri 1205 FF
> Hudburi, Hutbyr' 13th Dunst
> Utbiri 13th Dunst
> Obury 1826 B

This manor (v. burh) could only have been called Up on the lucus a non lucendo principle, for it is on some of the lowest ground in the parish. It does however lie in a remote corner of the parish and might well be described as the 'out manor.'

On the other hand, the fact that five out of seven medieval forms begin with h should not be ignored. There is no OE authority for an OE pers. name *Hutta which would give the best origin of the 1174 form Hutteberia with double t, and trace of possible inflexion in the medial e, but a name Hutting which can only be a patronymic from such a name occurs in the 12th cent., and we have Huttesbutt in 1219 (FF) in Eynesbury (Hu), and all the forms of Upbury could readily be explained by derivation from an OE (æt) Huttan byrig. If this derivation is correct the history of the forms is that tb was assimilated to db or bb and the db was then dissimilated to pb regardless of the sense suggested by the new form.

Silsoe

SILSOE [silsə] 95 A 9

> Siuuilessou, Sewilessou 1086 DB
> Siuelisho 1175 P

Siuelesho, Syuelesho 1199 P *et passim* to 1330 Fine
Shiuelsho(u) 1242 Fees 890, 1316 Ipm
Siuelho 1276 *Ass*
Syuelso 1323 BM
Sevelesho 1419 IpmR
Sybylso 1479 IpmR, 1485 Ipm
Silshoo 1506 AD v
Sylsoo 1535 VE

'Sifel's hoh' or hill. The name *Sifel* is not on record, but it has an OGer cognate *Sibilo* (Förstemann PN) and is a diminutive of the same *Sif-* stem which occurs in the well-known *Sifeca*, mentioned in *Widsith* (219), the treacherous adviser of Ermanaric. The same name is found in Silsworth (Nth).

ACRE POND (6″)
    cf. *Pondfurlong* 1262 FF
    Self-explanatory.

FIELDEN (House), FIELDING (Fm)
    la Felde 1205 FF, 1286 Dunst
    Fieldon Farm 1826 B

Probably descriptive of the open country (*v.* feld) in contrast to the wooded country in the north of the parish. The *en* is probably due to an alternative dat. pl. form (cf. Flitton and Ion *supra* 148, 151), the *ing* being a variant vulgarism (cf. Wilden *supra* 66).

WREST (Park)
    Wrest 1185 P (p), 1421, 1430 IpmR
    Wrast(e) 1276 RH, 1308, 1324 Ipm, 1329 Cl, 1331 QW, 1342
        Cl, 1354 Ipm, 1388 Cl, 1389 IpmR, 1396 Cl, 1413 IpmR,
        1610 Speed
    Wrast or *Rest* c. 1750 Bowen

This name is difficult. OE *wrǣste*, 'delicate, noble,' which may form the base of the pers. name found in Wrestlingworth *supra* 112, seems unlikely here and Professor Ekwall suggests some other derivative of the stem of OE *wrīðan*, 'to twist.' Norw has (*v*)*reist*, 'ring made of withies.' An OE *wrǣst*, 'something twisted,' may have existed, possibly in such a sense as 'thicket.' ON (v)reistr is used of Midgarðsormr and an element vrēst may occur in Swed names of lakes (Hellquist, *Studien över de Svenska Sjönamnen*, 728). It may be added that the ground round Wrest is much broken, and 'twisted, contorted' might be apt in a purely topographical sense.

Stopsley

STOPSLEY 95 E 9

Stoppelee 1199 FF (p)
Stopeleg, *Stopelee* 1201 Cur, 1240 Ass
Stoppeleg 1202 Ass, 1235 Cl
Stoppesle(gh) 1207 FF, 1227, 1240 Ass, 1247 *Ass*, 1286 Dunst et passim to 1504 Ipm
Stopeslegh 1247 *Ass*
Stop(p)isle(y) 1262 BM, 1487 Ipm
Stopeleye 1276 *Ass*
Stopesleye 1276 *Ass*, 1337 Ch, 1381 Cl
Stopley Eliz ChancP

Skeat and Ekwall (PN in -*ing* 71) are doubtless right in taking this to be from a pers. name *Stopp* found also in *Stoppingas*, the name of an archaic *provincia* now included in Warwickshire (BCS 157) and in Stopham (Sx).

CRAWLEY GREEN

Craulea 1196 Whet i. 421

'Crow-clearing' *v.* crawe. Cf. Husborne Crawley *supra* 118.

GALLEY HILL (6")

Galowehill 1504 Ipm

Self-explanatory.

MIXESHILL

Mix(e)weye 1276 *Ass* (p), 1297 *SR* (p)

If this identification is correct, the first part of this name is *Mixwey* from OE *meox-weg*, 'dung-road,' presumably so called because manure was carted along it.

RAMRIDGE END

> *Ramrugg* 1227 Ass
> *Ramrigge* 1240 Ass (p)
> *Ramerugge* 1290 AD iv (p), 1297 *SR* (p)

'Ram-ridge' *v.* ramm, hrycg or 'raven-ridge,' *v.* hræfn.

SOMERIES

> *Somereis* al. *Somereys* 1504 Ipm

The last trace of the manor of Greathampstead Someries, held by the Somery family from 1309 onwards (cf. VCH ii. 364).

WARDEN HILL

> *Wardonhill* 1504 Ipm

A hill with a commanding view, overlooking the Icknield Way. Hence we may take it, like Warden *supra* 97, to be from OE weard-dun, 'watch-hill.'

Streatley

STREATLEY 95 C/D 8

> *Strætlea* c. 1053 (c. 1250) KCD 920
> *Stradlei, Strailli, Stradli* 1086 DB
> *Stradlega, Stradlee, Stradlegh* 1166 P, 1242 Fees 887, 1247, 1276, 1287 *Ass*
> *Stredlea, Stredle(y)* 1171 P, 1276, 1287 *Ass*
> *Stratleg(h), Stratle(y)* 1231 Cl, 1247 *Ass*, 1275 Ipm, 1276 *Ass*, 1291 NI, 1297 *SR*, 1301 Ipm, 1302 FA, 1307 Ipm, Cl, 1316, 1346 FA, 1535 VE
> *Stretleg, Stretleye, Stretle* 1231 Cl, 1287 *Ass* (*juxta Sewendon*), 1297 *SR*, 1390–2 CS, 1526 LS, 1766 J
> *Stradeleye* 1276 *Ass*
> *Strettle* 1305 *Merton* 400 (p)
> *Stretly* 1806 Lysons

'The clearing by the stræt.' This 'street,' leading from Luton to Bedford, has not yet been proved to be a Roman road, and the term is probably descriptive only of a made-up road in contrast to an old track-way.

SHARPENHOE

> *Scarpeho* 1197 FF
> *Serpenho* 1197 FF
> *Sarpenho* 1227, 1240 Ass, 1285 AD iv
> *Sherpenho* 1231 Cl
> *S(c)harpenho* 1232, 1241 FF *et passim*
> *Scarpenho* 1240 Ass
> *Syarpenhou* 1267 Ch
> *Shappenho* 1316 Ch

OE *scearpan-hoge* (dat.), 'sharp spur of land,' aptly descriptive of the abrupt and steep promontory at the foot of which the village lies. *v.* hoh. For such a name, cf. *Scharphull* in Grafham (Hu), 13th AD iii.

Sundon

SUNDON 95 D 8

> *Sunnandun* c. 1053 (c. 1250) KCD 920
> *Sonedone* 1086 DB, 1276 RH, 1286 Dunst
> *Sunendon, Sonnydon, Shonendon* 1247 *Ass*
> *Sunedon* 1247 *Ass*, 1286 Dunst
> *Sonendone* 1276 *Ass* (p), 1291 NI
> *Sonin(g)don, Sonyngdon* 1276, 1287 *Ass*, 1315, 1322 Ch, 1328
>   Ipm, 1330 AD i, 1339 Ipm, 1372 Cl, 1373 IpmR
> *Sunyngdon* 1287 *Ass*
> *Sundon* 1390–2 CS
> *Sondon* 1489, 1499 Ipm
> *Soundon* 1535 VE

'Sunna's hill' *v.* dun. This pers. name is not known apart from place-names, but is presumably a shortened form of one of the OE names in *Sun-*, e.g. *Suneman*. It may be inferred as a name from the present example and from Sunbury (Mx), *Sunnanbyrig* BCS 1063, Sonning (Berks), earlier *Sunninges*. For the element *Sun-* in OE pers. names cf. Somersham *infra* 222.

IX. CLIFTON HUNDRED

Cliftone, Clistone 1086 DB

The meeting-place of the Hundred was presumably at, or near, Clifton, but the exact site is unknown. Clifton itself lies on the northern edge of the Hundred.

O

Arlesey

ARLESEY 95 A 12

Alricheseia 1062 (12th) KCD 813, 1086 DB

Alrichesei(e), *-eye* 1086 DB, 1247 *Ass*, 1251 BM, 1276 *Ass*,
 1306, 1317 Ch, 1322 Cl, Ipm, 1325 Cl, 1331 Fine, 1340 FA,
 1349 Cl, 1359 Ipm, 1386 Cl

Alricesei 1086 DB

Ailricheseia, *Aylricheseye* 1202 Ass, 1206 BM, 1224 FF, 1227
 Ch, 1242 Fees 882, 1247 *Ass*, 1253 Ch, 1254 BM, 1316 FA

Auricheseye 1220 LS, Hy 3 BM, 1287 *Ass*

Aillrikesheye 1227 Ass

Eylricheseye 1227 Ass

Eluricheseye 1227 Ass

Aluricheseya 1247 *Ass*

Hawricheseia 1255 BM

Alverycheseye 1270 Ch

Alrecheseye 1302 FA, 1331 Fine, 1350 Cl, 1438 AD iv

Alfricheseye 1307 *Ass*

Arlicheseye, *Arlycheseye* 1307 Ipm, 1402 BM, 1420, 1443
 IpmR

Arlechay 1386 BM

Arlechesey 1438 AD iv

Arlesey 1492 Ipm

Arseley al. *Alsey* 16th BHRS viii. 140

'Aelfric's well-watered land,' eg being used in its wider
sense. Skeat suggested that the pers. name was *Aeðelric*, but
forms unknown to him disprove this suggestion, though some
of the earlier forms in *Ail-* look as though early confusion with
that name took place[1]. As a rule, after the Conquest, names in
Aelf- and names in *Ael-* from *Aeðel-* are kept carefully apart.
Confusion was however possible. The same person e.g. is
described as *Hugo filius Alwini* in an original charter (Hy ii),
Hugo filius Ailwini in the Pipe Roll for 1189, and *Hugo filius
Elfwine* in an Assize Roll of 1202.

ARLESEY BURY (6″)

 Arlesey Berry 1667 BHRS ii. 141

---

[1] There is a second Arlesey (lost) in Cople (Beds), for which we have the
form *Ailricheshei* (*Warden* 84).

Etonbury (6")

Eatonsberry 1667 BHRS ii. 141

These two names are relics of the two chief manors in Arlesey. *v.* burh. *Eton* as usual denotes island, or well-watered farm.

Campton

Campton 95 A 10

Chambeltone 1086 DB

Camelton, Kamelton c. 1150 BM, 1176, 1180, 1185 P, c. 1180–90 NLC, 1189–96 NLC, 1185 Rot Dom *et passim* to 1548 Pat

Kamerton 1152–8 NLC

Kametona 1155 NLC

Cameltun c. 1172–5 (13th) *Dunst*, 1180 P

Gamelton 1177 P, 1276, 1287 *Ass*, 1291 Tax

Gamelinton 1276 *Ass*

Cameletone 1316 FA

Camulton 1334 Pat

Cambelton 1356 Cl, Ipm

Gameleton 1393 IpmR

Camulton 1400 CS

Campton 1526 LS

Campton al. *Camelton* c. 1550 *Linc*

Cambleton 1610 Speed

Skeat was probably right in conjecturing that we have here to do with a lost river-name. The name would be that of the un-named stream which ultimately joins the Hiz and with it soon joins the Ivel. Skeat quotes the doubtful parallel of the Camel River (Co), famous in Arthurian story. There would seem also to have been a derivative form of this river-name ending in *-ar* or *-er*. The stream now known as Cam Brook, which flows into Wellow Brook just before it joins the Avon a few miles south of Bath, is called *Camelar* or *Cameler* in BCS 1073. On its banks are Camerton, DB *Camerlertone* (sic) and Cameley, DB *Camelei*. From this it is clear that there is ample justification for assuming a river-name *Camel* or *Camelar*. We may explain the Bedford *Kamer-* forms as due to the common Anglo-Norman confusion of *l* and *r* or perhaps still better as due to the original name having been *Camelerton*. The forms of Cameley (So), in which no trace of the *r* is ever found, show how early it could be lost. The forms with initial *g* are due to association with names like *Gamlingay* from OE *gamol* or may be simply examples of the common confusion in ME place-names of initial *c* and *g* (cf. IPN 114). The *b* of the Domesday form may be compared with a *b* in similar position in the Domesday form of Hamerton, Hunts (*infra* 242). Tamerton (D), which is certainly derived from the river Tamar, appears as *Tambretone* in DB.

Chicksands Priory

CHICKSANDS [tʃiksəndz], [tʃiksən] 84 J 16

    Chichesane 1086 DB
    Chikesham 1152–8, 1155 NLC
    Chi(c)chesant 1156, 1158 P
    Chichesand 1159 P
    Chikes(s)ant 1161 P, 1196 FF, 1203 Cur, 1237 Cl, 1244 FF
    Chichessand, Chichesham 1162 P
    Chiksond 1163–79 BHRS i. 118
    Chik(k)essand 1185 P, 1276 *Ass*
    Chikesand c. 1190 (c. 1230) *Warden* 13, 1198 Fees 10 *et passim*
    Chikes(s)ond, Chyk- 1202 Ass, 1220 LS, 1297 *SR*, 1310 Cl,
        1316 FA, 1325 Cl, 1346 FA, 1361, 1385 Cl
    Chikes(s)aunde 1227 Ass, 1232, 1236 FF, 1244 Cl, 1247 *Ass*,
        1285 Ch, 1302 FA, 1316 HMC Var iv
    Chiksaund 1227 Ass, 1273 Cl
    Chikesaunt 1240 Ass
    Chikesend 1242 Fees 887
    Chijkesond 1250 Fees 1180
    Chikesonden(e) 1287, 1307 *Ass*, c. 1370-1420 *Linc passim*
    Chiksanden 1287 *Ass*
    Chikesand Dene c. 1300 *Linc*
    Chikeshanden 1307 *Ass*
    Chikesaundene 1317 HMC Var iv
    Chiksand 1327 Cl
    Chik-, Chyksond 1359, 1386 Cl, 1400 CS
    Chixham 1388 Cl
    Chiksonden c. 1390-1400 *Linc* (five times)
    Chikessounde dene 1428 FA

Chyxsond 1457 Ipm
Chickson 1655 NQ i

The suffix is the ordinary word *sand*. The soil here is sand (VCH ii. 271). The first element is an OE name *Cicca* which occurs again in Chickney, Essex (DB *Ciccheneia*). A variant of this name, with palatalised second consonant, is found in Chicheley (PN Bk 33), and a Latinised form *Cichus* (Redw 38) is recorded. Here we have an unpalatalised form. The forms with suffixed -*en(e)* which appear towards the end of the 13th cent. are difficult. Probably they point to an alternative dat. pl. form in -*sanden* (OE *sandum*). Such forms seem peculiarly common in Beds, cf. Flitton, Ion, Fielden *supra* 148, 151, 162. The form with *Dene* in the Lincoln Registers might be a piece of folk-etymology, but it is unlikely in so conservative and highly formal a series of records as a Bishop's register. It is however possible that this and similar forms should be compared with the curious form *Wicumbedene* for High Wycombe found in the earliest Pipe Rolls of Henry ii. A medieval use of *dene*, of unexplained origin, to denote a district dependent upon or annexed to a place is not impossible. It seems indeed to have survived in the name Taunton *Dean* for the great manor of Taunton.

It is of course conceivable that there was a form *Chikesanddene* with the ordinary word *dene* (*v*. denu) suffixed, but there does not seem to be a sufficiently well-marked valley here to make such a new development likely, though in the Warden Cartulary (19 *b*) we have mention of a *Grenedene* in Chicksand. The whole name means 'Cicca's sands.'

APPLEY

Appeleia c. 1150 BM
Appele 1276 *Ass* (p), 1287 *Ass*
Appelaya, boscus de Hy 2 BHRS i. 118
'Apple-tree clearing' *v*. æppel, leah.

Clifton

CLIFTON 84 J 11

Cliftune 944–6 (c. 1250) BCS 812
Cliftone, Clistone 1086 DB
Clyfton juxta Schefford c. 1350 *Linc*

Self-explanatory, though the name is a good illustration of the fact that in OE place-nomenclature it does not take much to make a 'cliff.'

THE GRANGE

Grangefeld 1276 *Ass* (p)

Self-explanatory.

HOO HILL

le Ho 1242 Fees 869 (p)

One more example of this favourite Beds p.n. element. *v.* hoh.

Henlow

HENLOW 95 A 11/12

Haneslauue, Hanslau, Hanslaue, Haneslau 1086 DB

Hanelawe Hy 2 (1261) Ch, 14th Gest St Alb

Hanlaga 1202 Ass

Hennelawe 1207 FF, 1287 *Ass*

Hanlawe 1220 LS

Henlawe 1227 Ch, Ass, 1242 Fees 867, 1247 *Ass*, 1253 FF, 1276 *Ass*, RH, 1299 Ipm, 1399 Ch

Anelawe 1227 Ass

Hanlowe 1276 *Ass*

Hannelowe 1282 Abbr

Henlowe 1302 FA *et passim*

Apart from the Domesday forms, with an inflexional *s* for which there is no support, all the spellings of this name point to derivation from OE *hænna-hlāw*, 'fowls' hill' (*v.* hlaw), with the same development of ME forms in *Han-* as well as in *Hen-* from OE *æn* which we have in Fancott and Dane End *supra* 137, 153. In these last two names, however, the *an-* forms have prevailed. If this derivation is correct, the name is exceptional, for OE *hlaw* is usually preceded either by an adjective (cf. Goldenlow *supra* 120) or, most commonly, by a pers. name.

Meppershall

MEPPERSHALL [mepʃul] 95 A 10

Malpertesselle, Maperteshale 1086 DB

Maperteshala Steph (1313) Ch, 1190 P, 1198 Fees 10

Meperteshale 1200, 1206 FF *et passim* to 1490 Ipm
Mainpardeshal 1203 FF
Maperteshale Hy 2 (1255), 1202 Ass, 1220 LS, 1224 Bract, 1247 *Ass*, 1255 Ch
Mepardeshale 1227 Ass, 13th AD ii
Mauperteshal' 1233 Bract
Meperdeshale 1247 *Ass*, 1244–52, 1285 Ch, 1302 FA, 1445 AD iii
Meper(e)shale 1316 FA, 1369 Fine
Mepersale 1331 Ipm
Meparteshale 1347 Pat
Mapartysshall, Meppertyshall 1494 Ipm
Mepersall 1526 LS, Eliz ChancP
Mepsall 1610 Speed, 1635 NQ i
Meppershall or *Mepshall* 1806 Lysons

'Mathalperhts's nook or corner of land' *v.* healh. The pers. name in question is the OHG form of an OE name *Mæðelbeorht*, supposing such to have existed. *Mæðel-* names are very rare in OE, *Mæthelgār*, from which Maugersbury (Gl) is derived, is the clearest instance, and *Mæðelbeorht* has not yet been found. Förstemann gives several examples of *Madalper(a)ht*, and it is probable that here we have to do with a late settler or, more probably, some king's thegn of continental origin who became possessed of this piece of land. On the other hand, the name was clearly recognised as the equivalent of an English name in *Mæðel-*, as thus only can we account for the development of *Mep-* forms, which ultimately triumphed over the *Map-* ones which would be the normal development of a name in *Mathal-*. This does indeed suggest the possibility that the name was really OE *Mæðelbeorht* and that the *p* was due to an unconscious assimilation of this extremely rare English name to its continental cognate.

Hoo (Fm)
   le Hoo 1393 AD iv
   v. hoh.

POLEHANGER *olim* [puliŋgə]
   Polehangre 1086 DB, 1247 *Ass*
   Pullehang' 1198 Fees 10

Polhangre 1220 LS (p), 1276 *Ass* (p)
Pulehangre 1227 Ass
Pulhangre 1247 *Ass*, 1291 Tax, 1402 BHRS i. 102
Pollehangre 1276 *Ass* (p)
Pullanger 1445 AD iv, 1461, 1483 IpmR, 1490 Ipm
Pullenger Eliz ChancP

It is natural to translate this name as 'pool-wood,' *v.* pol, hangra. The ground slopes down fairly steeply to the farm. If this explanation is correct the pool must be one in the Ivel near the place, for there is no other neighbouring pool[1]. In view of this difficulty, the possibility should not be ignored that the first element here represents a pers. name. The existence of *Pul, Pol,* as the base of such a name is proved by the diminutive form *Pulloc* or *Polloc* which occurs in Pulloxhill, cf. also Pulworthy (D), earlier *Poleworthi*. It is worth noting that half the forms earlier than 1300 show an *e* between the *l* and the *h,* which may be the remnant of an inflexional *an*. Moreover, if the name were a compound of *pol* and *hangra*, the forms of 1198 and 1276 would be at least abnormal. They suggest strongly derivation from a weak pers. name with a double consonant, *Pulla* or *Polla.*

WOODHALL (Fm)
   cf. *atte Wode* 1445 AD iv (p)
   Self-explanatory.

Shefford

SHEFFORD 84 J 10/11
   Sepford 1220 LS
   Shipford 1229 Cl
   Shepford 1247 FF
   Sefford 1247 *Ass*, 1251 BM
   S(c)hefford 1262, 1271 FF
   Che(f)ford 1276 *Ass*, 1361 Cl
   Sheford 1276 *Ass*, 1297 Ipm
   Schyford, Chyford 1287 QW
   Schepeford 1307 *Ass*
'Sheep-ford' *v.* sceap, ford.

[1] *ex inf.* the Rev. R. Isherwood.

SHEFFORD BRIDGE

pons ad caput villae de Shefford 1276 *Ass*
Skegfordbregge 1287 *Ass*

Self-explanatory. The 1287 form shows a curious effort at
Scandinavianising the name.

Shillington

SHILLINGTON 95 B 10

Scytlingedune 1060 (14th) KCD 809
Sethlindone 1086 DB
Scetlingedon 1202 *P*
Scetlingdon 1207 *P*
Sutlingedon 1220 LS
Setlingdon 1221 FF
S(c)hitlingdon 1222 FF, 1276 *Ass*
Sut(h)lingdon 1227 Ass, 1265 BM, 1266 ADi, 1286 Dunst
Sytlingdon 1227 Ass
Schutlingdon 1236 FF
Shutlingedon 1239 FF
Sitlingedon 1240 Ass
Sutlindon 1242 Fees 869
S(c)hittlingdon 1247 *Ass*, FF (p)
S(c)hitlingdon 1251 Ch, 1256, 1258, 1261 FF, 1272, 1276 *Ass*,
    1294 BM, 1388 Cl, 1397 AD i, 1415 BM, 1504 Ipm
Schitlingedon 1253 BM
Schetlindon 1262 FF
Sitelyn(g)don 1276 *Ass*
Shittelington 1276 *Ass*
S(c)hutlyngdon, -ingdon 1276 *Ass*, 1291 NI, 1297 *SR*, 1316
    FA, 1332 Ipm, 1333 Cl, 1338 AD i, 1351, 1368, 1381 Cl
S(c)helyngdon 1287 *Ass*, 1400 CS
Schettlingdon 1287 *Ass*
Schudlyngton 1287 *Ass*
S(c)hytlington, Shitlington 1287 *Ass*, 1368 Cl, Eliz ChancP,
    1780, 1820, 1830 Jury
Shutlyngdene 1302, 1316 FA
Shutlyndon 1327 Cl
Shutlyngton 1367 Orig, 1372 Cl

Shetlington, Shetlyngton 1526 LS, 1535 VE
Shedlington 1675 Ogilby
Shidlington 1675 Ogilby, 1780 Jury
Shilindon 1780 Jury

'Hill of Scyttel's people' *v.* dun, inga. The existence of this pers. name is made certain by Shitlington (Y, Nb), and Sheepstor and Sheepsbyre (D), earlier *Schetelestorre, Shitelesbere.* Cf. more fully PN NbDu 178[1].

APSLEY END

Aspele 1230 Bract, 1253 Ch

The same as Aspley *supra* 113.

BURY END

Shitlington Bury 1409 AD vi

The 'manor' farm, *v.* burh.

CHIBLEY (Fm)

Chubele 1255 (c. 1350) Rams (p)
Chybbele 1301 *Ct*

'Ceobba's clearing' (*v.* leah). For this pers. name we may compare Chibley (So) which is *Chubbeleye* (1305 Ipm) and Chubworthy in the same county, DB *Cibewrde,* Ipm 1285 *Chubbeworthe.*

FEAKS WELL (lost)

Fageswell 1218 ClR, 1276 *Ass,* 1303 *KF* (p)
Faggeswell 1247 *Ass,* 1287 *Ass* (p)
Fagewell 1276 *Ass*
Faukewell 1291 AD iv (p)
Feaks Well 1766 J

Professor Ekwall suggests the parallel of *Fackeswell* in London in a Fine of 1291. We may note also Faxton (Nth), 1166 *Fachestuna.* There is evidence for an OE pers. name *Facca* in Faccombe (Ha), *Faccancumb* Thorpe 534, and in *Faccanleah* (BCS 1232), allied to the pers. name *Fecca* found in Feckenham (Wo). It is found as *Fache* in the Selby Cartulary (i. 205, 224) in the 13th cent. In its strong form *Fæcc* a gen. *Facces* may at times

---

[1] Shillington was divided between Flitt and Clifton Hundreds. Aspley, Bury and Pegsdon were in the former.

have become *Fagges* (*v.* Cople, Moggerhanger, Cogswell *supra* 89, 91, 98). The later phonological development is wholly irregular.

HANSCOMBE END

Hanescamp 1222 FF (p)
Hameschaump 1227 Ass (p)
Hanescompe 1255 (c. 1350) Rams, 1287 *Ass* (p), 1297 *SR* (p)

The second element is the somewhat rare OE camp, 'open country,' which nearly always appears now as *combe* on our maps. Professor Ekwall suggests the possibility of an OE pers. name *Hān* (cf. *hān*, 'hone') from which we seem to have a derivative patronymic form in Honing (Nf). Cf. also Hannington (Nth) with shortening of *ā* in the trisyllable.

HOLWELLBURY

Helewell 1195 P
Holewell 1197 FF (p)
Magna, Parva Holewell 1242 Fees 869
Parva Hollewell 1276 *Ass*
Holewellebury 1420 Ipm
Hollowell Eliz ChancP

'Holwell manor' *v.* burh.

Some of the above forms refer to Holwell itself, which is in Herts.

PEGSDON [pegsən]

Pechesdone 1086 DB
Pekesdene 1114–30 (c. 1350) Rams, 1228 FF, 1247 *Ass*, 1255 (c. 1350) Rams, 1264 AD iii, 1265 BM, 1266 AD i, 1276 *Ass* (p), 1287 *Ass*, 1290 Cl, 1297 SR (p)
Pekedene 1205 FF
Peckesden 1227 Ass, 1287 *Ass*
Pikelesdene 1227 Ass
Pekelesdene 1227 Ass
Pakesden 1227 Ass
Pachesdena 1230 Bract
Pekesdon 1240 Ass, 1247 *Ass*, 1276 *Ass* (p), 1302 FA
Peckesden 1287 *Ass*

Pexden 1350 BM
Pegson c. 1750 Bowen, 1766 J

Professor Ekwall has contributed the following note upon this name:

The first element is identical with the p.n. *Pek* found as surname of a tenant in Pegsdon (*Miles de Pek* Rams Cart i. 471) as noted by Skeat. With this *Pek* may be compared *Pec*, the name of a place in Ganton (Y), also in the compounds *Pekespit*, *Pekesbru* (all in a 13th cent. ch. in the Bridlington Cart), and also the well-known *Peak* (Db). The latter appears in OE sources as *Peac* (cf. *Peaclond* ASC 924, *Pecsætan* Trib Hid). Curiously enough there is an early compound in which also this name appears in the gen. form, viz. Peak's Arse (Peak Cavern) *Pechesers* DB. *v.* Addenda. There is good reason to believe that the name in all three cases originally denoted a hill. As for Peak there can hardly be any doubt. Henry of Huntingdon mentions the mountain (mons) called *Pec*. The name appears to have been originally applied to Castle Hill at Castleton. *Pec* (Y) is preceded by *supra*, and *Pekesbru* must contain OE *brū*, 'brow,' here in the sense 'brow of a hill' (perhaps the earliest known instance of this meaning). Ganton is on the slope of a considerable hill, which may have been known as *Pek*. East and West Peek (D), DB *Pech*, are on a hill rising to 569 ft. Further we may note that Pegsdon is at the foot of a very well-marked and steep hill. If this suggestion is correct, Pegsdon would mean '*Pek* hill' and be a formation analogous to *Andredesweald* and the like. It should be added that the identity of the first element of Pegsdon with Peak is suggested already by Johnston, *PN of England*.

If the three names have a common origin, the base must be OE *pēac*, which may be taken to be an old hill-name. We expect the name of a hill such as the Peak to have a pre-English name, but the form of the name with its diphthong *ēa* rather suggests Germanic origin, and if the three *peks* are etymologically identical, Germanic origin becomes still more probable. There is a well-known Germanic stem *puk*, *pauk*, found in Du *pôk*, 'a dagger,' Swed *påk*, 'a cudgel,' Engl *poke*, *pock*, *Puck*, etc., Norw dial. *pauk*, 'a stick, a little boy,' etc. Words belonging to this group often denote a rounded object, a thick-set figure and the like. A meaning 'knoll, hill' would easily develop from that. The exact counterpart of OE *pēac* would be the Norw *pauk*, though the meaning is rather different.

Upton End

Uppennende 1255 (c. 1350) Rams 465, 1294 *Ct* (BM) (p)
Opton 1276 *Ass*

This apparently simple name is not easy. *Uppenende* would seem to be simply a compound of OE *uppan*, 'upon,' and ende, and one could conceive of a man being called Ricardus *Uppenende*, 'Richard at the end of the parish,' but Ricardus *de Uppenende* is a little strange. In the first example *Uppenende* is given as the name of a hamlet in Shillington. One can only suggest that the hamlet was at first called 'At the end' (OE *uppan ende*) and that later the name was corrupted to a fresh and more usual form. Association with *up* is out of the question as this 'end' is distinctly one of the lower 'ends' of Shillington parish.

Woodmer End

Wodemenende 1255 (c. 1350) Rams (p)
Wodemanende 1297 *SR* (p)
Wodemannande 1300 *Hunts Ct Rolls* (PRO) (p)
Woodmore End 1766 J

'Woodman's or woodmen's end' *v.* ende.

Upper Stondon

Upper Stondon 95 A 11

Standone 1086 DB *et passim* to 1712 BM
Staundon 1199 FF, 1276, 1287 *Ass*, 1291 Tax, AD iv
Stondon(e) - 1247 *Ass*, 1252 FF, 1276, 1297 *Ass*, 1302, 1316, 1346 FA, 1388 Cl, 1390–2 CS, 1428 FA
Great Stondon 1320 Ipm
Overstondon 1504 Ipm

'Stony-hill' *v.* stan, dun. *Over* is the regular earlier epithet for places now called *Upper*.

Stotfold

STOTFOLD 95 A 13

Stodfald 1007 Crawf 11, 1193 P, 1202 FF, 1203 Cur (p),
1227 Ass
Stotfalt 1086 DB
Stotfald 1198 FF, Hy 3 (1317) Ch
Stotfold 1199 FF, 1232, 1242 FF, 1287 *Ass*, 1302 FA *et
passim*
Stotfaud c. 1200 BM, 1227 Ass, 1276 *Ass*
Stofaude 1202 Ass, 1236 FF
Stodfold 1227 Ass, 1244 FF
Stodfauld 1227 Ass
Stodfaud 1247 *Ass*
Stotfeld, Stotfeud 1276 *Ass*
Stotfolt 1287 *Ass*
Stottesfold or *Stotfold* 1343 Ipm
Statfold(e) Eliz ChancP, 1780 Jury

'Stud-enclosure,' *v.* stodfald, but, for a possible further
significance of the name, *v.* IPN 151.

WILLBURY HILL FARM

Wiligbyrig 1007 Crawf 11

The *bury* here is clearly the camp on Willbury Hill, the whole
name being 'willow-camp.' *v.* welig, wylig.

Kensworth

(Formerly in Hertfordshire)

KENSWORTH 95 F/G 7

Canesworde 1086 DB
Kenesword 1131–3 (13th) *Dunst*, 1286 Dunst
Keneswurda 1168 P (p)
Keneswurth R i (1227) Ch, 1247 *Ass*, 1286 Dunst *et passim*
Keneswrthe 1227 Ass
Kemeswurth 1247 *Ass*
Kenesworth 1286 Dunst, 1299, 1302 AD iii, 1402 AD i

Kensworth 1375, 1416 AD vi
Kennesworth 1431 AD i, 1434 AD iii, 1435 AD ii

'Cen's enclosure' *v.* worð. The pers. name *Cēn* or *Cǣn* is on record in OE and is a pet-form of one of the numerous names in *Cǣn-*.

SLOUGH WOOD (6″)

de la Sclo 1297 *SR* (p)

Self-explanatory.

HUNTINGDONSHIRE

I. NORMANCROSS HUNDRED

Norðmannescros 963 E (12th) ASC
Normannes cros 963–84 (c. 1200) BCS 1128, Stephen (c. 1300–25) *Thorney* 9 *b*
Normanescros 1086 DB
Normanecros 1086 DB
Normancros Wm 2 (c. 1300–25) *Thorney* 8 *a*, 1255 *For*, 1292 Fine
Normannecros 1160 P, 1227 *Ass*
Normannescros Stephen (1314) Ch

Norman Cross stood on Ermine Street where it is crossed by the road from Yaxley to Folksworth. It stands roughly at the centre of the group of parishes which form the Hundred. The name is interesting for it, like that of Toseland Hundred, is clearly of Scandinavian origin. The first element is not the word *Norman* but the older *Norðman* applied by the Anglo-Saxons to those Scandinavians who came from Norway, and used also as a pers. name descriptive of some one from that country. It is clear that it is in the latter sense that the word is used here, but who the particular *Norðman* was who thus gave his name to the cross we do not know. *cros* is itself a Norse loan-word (*v.* EPN *s.v.*), and it is interesting to note that in the only hundred-names in which the meeting-place is, by the very name, indicated as the site of a cross, we have this Scandinavian loan-word and not the English rod or mæl, though it is just possible that *tree*-names which are fairly common as hundred-names may sometimes refer to a cross rather than to a living tree. Other examples of *cross* in hundred-names are Brothercross and Giltcross in Norfolk and Staincross and Buckrose in Yorkshire. Presumably there was some earlier Anglo-Saxon name for this Hundred, now lost.

Alwalton

ALWALTON [ælətən] 74 A 10
  Aeþelwoldingtun 955 (c. 1200) BCS 909
  Alwoltune 1086 DB
  Alevoltone c. 1125 (c. 1200) Lib Niger de St Petroburg

Aðelwoltun' 1158 P
Alwoldton 1176 P (p)
Alewalton 1189 (1332) Ch, 1227 Ch, *Ass*, 1292 Orig, 1300 Ch
Ayllewolton 1245 *For*
Alwalton 1268 Ch, 1316, 1428 FA
Aylwalton 1270 *Ass* (p), 1286 *Ass*, QW, *For*
Alwaldon 1292 Fine, Orig
Alerton 1610 Speed
Allerton 1675 Ogilby
Alwalton or *Allerton* c. 1750 Bowen

'Æðelweald-farm' *v.* ingtun. The form found in DB and later is not to be regarded as a reduction of the fuller OE form but rather as an alternative name in which the pers. name was directly prefixed to tun without any use of either the *-ing-* element or, what is even more rare, any use of the genitival suffix in *-es*.

Caldecote

CALDECOTE 74 C 10

Caldecote 1086 DB *et passim*, 1286 *Ass* (*cum Denton*)
Caudecote 1248 *For*
Coldecote 1301 Cl
Calcote 1504 Ipm, 1514 LP
Calcot(t) 1526 LS, 1610 Speed
Caldecote al. *Calcott* 1551, 1570 FF
Cawcott 1576 Saxton

'Cold cottages' *v.* cald, cot.

Chesterton

CHESTERTON 74 A 10

Ceastertuninga gemærie (sic) 955 (c. 1200) BCS 909
Cestretuna 1086 DB, 1184 BM
Cestreton, Cesterton 1192 BM, 1217 Pat, 1227 *Ass*, 1257 FF, Ch, 1260 *Ass*, 1286 Cl, *Ass*, 1290 Ipm, 1303 FA, 1304 AD v, 1316 FA, 1326 FF
Chesterton, Chestreton 1345 Ipm, Fine, 1362 Cl, 1428 FA
Chasterton 1355 FF, 1362 Cl

'Farm by the ceaster,' the reference here being clearly to the Roman camp at Durobrivæ. *v.* tun.

P

Conington

CONINGTON [kʌniŋtən] 74 D 11

    æt Cunictune 957 BCS 1003
    Cunintone c. 1030 (c. 1300–25) *Thorney* 79, 1227 *Ass*
    Coninctune 1086 DB
    Cunitona c. 1180 BM
    Cunnington 1214 *Fine*, 1662 Fuller
    Cunington, -yng- 1227 *Ass*, 1237 Cl, 1296 FF
    Coniton, -yt- 1235 Cl, E 1 BM, 1303 FA, 1328 FF, 1330 Cl,
       Ipm
    Cunytun 1236 Cl
    Conyngton, Conington E 1 BM, 1290 Cl, 1317 FA, 1318, 1320,
       1350 Cl *et passim*
    Conynton 1283, 1294 Cl, 1323 Ch, 1353 Cl
    Connyngton 1585 FF
    Conington or *Cunnington* c. 1750 Bowen

The history of this name is not certain. On the whole the probability is that here and also in Conington (C), *Cunningtun* BCS 1306[1], we have a Scandinavianising of OE *Cyning-tun* or *Cyne-tun*, 'king-farm' or 'royal-farm,' under the influence of ON *konungr*. The more usual form both in OE and ON is with a genitival suffix, but we do find in Old English *Cyngtun* (BCS 1234) now Kineton (Wa) and the same form (KCD 570) for Kington (Wo), while Kingston Bagpuize (Berks) appears without an -*es* in *Cingtuninga gemære* (BCS 1047). Skeat's suggestion of derivation from an OE pers. name *Cun(n)a* is, on the whole, less probable.

BRUCE'S CASTLE FARM

The Bruce family held Conington manor (cf. FF 1321).

CONINGTON BROOK (6″)

    Todbrok 1411 *Ct*
    Totebroke 1488 *Ct* (T. S. H.)[2]

---

[1] It is perhaps worthy of note that the *Textus Roffensis*, as quoted in *Proc. Soc. Antiq.* (2nd series), iii. 49, gives an OE form *cunigtun* for this name, which affords a good parallel to *Cunictun*.
[2] The initials T. S. H. indicate our indebtedness to the kindness of Mr T. S. Heathcote.

'Fox-brook,' if we take the first element to be ME *tod*, 'fox,' though there are difficulties in doing so as that word is distinctively a North Country word, at any rate in early usage. If *Tote-* is the correct form, with later voicing of *t* to *d* before *b*, then the first element would seem to be the pers. name *Tota*.

CONINGTON FEN

mariscus de Conyngton 1253 *For*

ETERNITY HALL

'From a nickname given to Edward Smith who lived there from 1845 to 1896' (T. S. H.).

OUTERNESS WOOD (6")

Uttyrness 1483 *Ct*
Le Uttyrness 1498 *Ct* (T. S. H.)
Utternesse Estate Map c. 1600, 1772 *Rent Book* (T. S. H.)

'Outer-ness' *v.* næss, ME *utter* being the earlier comparative form of *out*, now replaced by *outer*. The reference is to a clay promontory jutting out into the Fens. It is probably the 'outer' ness in contrast to *La Nesse* mentioned in 1279 (RH), which Mr Heathcote identifies with the site of Bruce's Castle.

Denton

DENTON 74 C 11

Dentun 972–92 (c. 1200) BCS 1130
Dentone 1086 DB *et passim*, 1327 SR (*cum Caldecote*), c. 1400 Linc (*juxta Stilton*), c. 1415 ib. (*prope Jakesle*)

Situation and form alike make it clear that this is OE denu-tun, 'valley farm,' rather than 'farm of the Danes.'

Elton

ELTON 74 A 9

Æþelingtun, Æilintun 972–92 (12th) BCS 1130
Adelintune 1086 DB
Aethelyngtone 1123–36 (c. 1350) Rams
Ailincton Hy 2 BM
Adelington 1199 Cur
Ail(l)inton 1207 P, 1209 *For*

Ayl(l)ington, Aylyngton 1215 FF, 1244 (c. 1350) Rams, 1260
   Ass, 1267–85 AD iv, 1285 FF, 1291 NI, 13th AD i, 1304
   AD v, 1316 FA, 1327 Cl, 1355 FF
Eylinton 1227 *Ass*
Elinton 1227 *Ass*
Aylingeton 1248 FF
Athelinton, Ethelintun 1253 BM
Eylington 1260 *Ass*
Alyngton 1288, 1295 FF, 1392 BM, 1501 Ipm
Alynton 1303 Orig
Aylyngton al. *Aileton* 1517 BM
Aylton, Ailton 1535 VE, 1539, 1549 BM, 1593 FF

'Æðel's farm' *v.* ingtun, *Æþel* being a shortened form of one
of the numerous OE pers. names in *Æðel-*.

BOTOLPH GREEN (6″)

Bottle Green 1766 J, 1826 G
Pottle Green 1836 O

If we could recover the early forms of this name we should
probably find that its history was the same as that of Botolph
Bridge, viz. that it was named first after St Botolph, then by a
process of popular corruption became Bottle Green, and that
finally some polite antiquarian or map-maker restored the full
name of the saint.

DERNFORD (lost)

Derneforde 1164 BM, 1279 RH
water of Dernford 1589 FF

'Hidden ford' (*v.* dierne, ford). So effectually is it hidden that
none is marked on the O.S. map, but it may be presumed that
it was where a foot-path leads away from the river-bank just
across the Nen from Elton village.

THE GREEN (6″)

ad Grenam 1303 *SR* (p)
Self-explanatory.

OVER END

cf. *Netherhende* 1218 (c. 1350) Rams
Neyerhende 1279 RH

No early form for the *Over* or Upper End has been noted, but it clearly stood in contrast to a *Nether* End. Over End is at the top end of the village.

Farcet

FARCET [fæsət] 74 A 12

Fearresheafde (dat.) 955–9 (c. 1300–25) *Thorney* 6 *b*, c. 1000 *Cragg*

Farresheafde (dat.) 963–84 (c. 1200) BCS 1128

Faresheued 963–84 (c. 1200) BCS 1128, 1278 *Ass*, 1316 FA, 1327 *SR* (p), 1353 Ch

Farresheued c. 1150 (c. 1300–25) *Thorney* 168 *b*

Fayresheued 1260 *Ass*

Farisheued 1260 *Ass*

Farsheued 1260 *Ass*, 1279 RH, 1327 *SR*

Fersheued 1260 *Ass*

Fasset(t) 1526 LS, 1576 Saxton, 1595 FF

'Bull's head' *v.* fearr, heafod. For the possible significance of names of this type, *v.* Swineshead *supra* 21.

FARCET BRIDGE and FEN

Faresheved Brygg, Fen 1279 RH

KING'S DELPH GATE

Cynges dælf 963 E (c. 1200) ASC

Cnoutes delfes kynges 1052–5 (c. 1350) Rams

Kyngesdelf 1286 (c. 1350) Rams

This delf is a channel, thus described by Dugdale (*History of Embanking*, 2nd ed., 363): 'About two miles distant from the north-east side of Wittlesey Mere, there is a memorable channel cut through the body of the Fen, extending itself from near Ramsey to Peterborough, and is called King's delph. The common tradition is that King Canutus, or his Queen, being in some peril, in their passage from Ramsey to Peterborough, by reason of the boisterousness of the waves upon Wittlesey Mere, caused this ditch to be first made,' in support of which he quotes a passage from the pseudo-Matthew of Westminster. This tradition is confirmed by the form quoted above from the Ramsey Cartulary which similarly identifies the king with Canute. Dugdale noted the inconsistency of this tradition with the mention of King's Delph in a Peterborough charter ascribed to King Edgar, but seeing that all these early Peterborough charters are notorious forgeries we can keep Canute and believe him to have been the promoter of this piece of early fen-engineering.

MILBY

Myleby 1279 RH (*pratum*)

This was originally only the name of some meadow-land, and it is not likely that the second element in it is the common Danish *by*. From the map it would appear that the name is applied to an area rather than to a particular spot. It is situated just where the Nen makes a well-marked bend in its course and perhaps the second element is OE *byge* (cf. Beeston *supra* 108). What the first element is one cannot say.

RAWERHOLT (lost)

silua Ragreholt a. 1022 (c. 1200) KCD 733
Reyereholt 1279 RH
Rawereholt c. 1350 Rams

This wood lay between Whittlesey Mere and King's Delph. It is clearly from OE *hragra*, 'heron,' and holt, hence 'heron-wood[1].' Cf. Rawreth (Ess) which is from OE *hragra-hyð*, 'heron-hithe.'

Fletton

OLD FLETTON 64 J 12

Fletun 1086 DB
Flettuna c. 1125 (c. 1200) Lib Niger de St Petroburg
Fletton 1227 *Ass* (*et passim*)
Flutone 1316 FA

'Farm on the *fleet* or stream' v. fleot. v. Addenda.

Folksworth

FOLKSWORTH *olim* [fɔxwə·θ] 74 B 10

Folchesworde 1086 DB
Fulkewwurþe c. 1150 (c. 1300–25) *Thorney* 169 a
Fulkesw(o)rthe 1152–8 NLC, 1220 Fees 334, 1220, 1240 *Ass*, 1248 For, 1276 *Ass*, 1316 FA

[1] Stevenson MSS.

Fuchowurda 1155 NLC
Fukessord 1160–5 NLC
Fulcheswurda 1167 P, 1181 P (Chanc Roll)
Folkesw(o)rth(e)   1185 (c. 1200) *Templars*, 1253 FF, 1260,
   1270 *Ass*, 1286 *For*, 1292 Orig, 1322 Ipm *et passim*
Fukeswrth 1201 Cur (p)
Fukeswurth 1227 *Ass* (p)
Fakewurth 1227 *Ass*
Fokesworth 1239 FF
Foukewrth 1253 *For*
Falkesworth 1267–85 AD iv (p)
Fukewurth 1270 *Ass*
Foukeswurth 1272 FF
Folkewurth 1276 RH
Fakesworth 1314, 1315 Ipm
Felkesworth 1348 Cl
Foxworth 1526 LS

'Folc's enclosure' *v.* worþ. *Folc* is not on independent record
in OE but is a regular shortened form of OE names in *Folc-* and
is found both in its strong and weak form in *folcanstan* (BCS 408)
and *folces stan* (ib. 813) for Folkestone (K). The numerous *Fulk-*
forms are to be accounted for by the influence of the common
French name *Fulk*, itself ultimately of the same origin.

Glatton

GLATTON 74 D 11

glædtuninga weg 957 BCS 1003
Glatune 1086 DB
Glattun 1158 P, c. 1200 BM, 1217 Pat
Glatton 1167 P *et passim*, 1316 FA (*cum Holme*)
Gletton 1260 *Ass*

'Cheerful' or 'pleasant' farm. For the use of the adj. *glæd* in
place-names, cf. a possible instance in Nares Gladley *supra* 125.
The *weg* or road of the men of Glatton is probably the old cart-
track just to the south of Glatton village, along which the
bounds of Conington parish still run as they did in the days of
the charter from which the extract is taken.

Haddon

HADDON 74 B 10

> *æt Haddedune* 951 (c. 1300–25) *Thorney* 6 *a*
> *Adone* 1086 DB
> *Haddune* c. 1150 (c. 1300–25) *Thorney*
> *Haddon* 1286 *Ass*, 1316 FA, 1327 *SR*, c. 1370 *Linc* (*juxta Yakesley*)

It is difficult to know how much stress should be laid on the *Hadde-* of the form in the Thorney Register. Probably it should be taken definitely into account and the first element be taken as the pers. name *Hædda*. Cf. *Headdandun* (KCD 544). Apart from this we should have taken it as OE hæð-dun, 'heath covered hill,' a very common form of p.n.

Holme

HOLME 74 C 12

> *Glatton cum Hulmo* 1167 P
> *Hulm(e)* 1217 Pat, 1224 FF
> *Holme* 1252 Ch *et passim*
> *Ulmo* (abl.) 1303 *KF*
> *Houlyn* 1312 Fine
> *Hulmus* 1319 FF
> *Home* 1526 LS

v. holmr. The name is descriptive of its water-surrounded site. The persistent *u*-forms, pointing to ODan *hulm* rather than *holm*, are noteworthy. For their significance, cf. IPN 60 and Ekwall PN La 244.

CHALDERBEACH FARM (6″)

> *Scælfremære, Scelfremære* a. 1022 (c. 1200) KCD 733
> *Chelfremerebeche* 1146 *Cott* vii. 3
> *Chelfrebecche, Cheluremerebeche, Chelwremere, Chelwremare* c. 1150 (c. 1350) Rams
> *Scelremere, Salderemere Breche, Saldermere* 1279 RH

This mere lay to the south of Whittlesey Mere, and Chalderbeach Farm must have been on its eastern shore. Its name was clearly *scealfra-mere*, i.e. 'mere of the diver-birds.' Later the *f* was lost between the *l* and *r* in the consonant group and was replaced by the epenthetic *d* which so commonly develops in ME between *l* and *r*. Confusion of initial *sch* and *ch* is fairly common and would in this case be assisted by the influence of the common ME *chaluer, cheluer*, gen. pl. of 'calf.' The *beach* is clearly the same element that is found in Wisbech, Land-beach and Waterbeach (C) and refers to the position of the farm on the edge of the mere. The word may be the same as OE *bæc*, *bece*, 'stream,' with transference of sense from the stream to its bank, but this is uncertain (*v.* Weekley, *Etymological Dictionary*, s.v.). *v.* Addenda.

DOCK A FALT (lost)

aqua Falet a. 1022 (c. 1200) KCD 733
Falt, nouam Falt super Witlesmare 1224 *FF*
Falthal' 1227 *Ass*
Faltestub 1286 *Ass*
Dock a Falt c. 1750 Bodger's *Map of Whittlesea Mere*

This lost name of a spot on the east side of Whittlesey Mere clearly contains the name of a tributary stream of that mere, named in the Saxon list of its bounds. Professor Ekwall calls attention to *Falete* (1251 Ch), now Fauld (St), DB *Felede*, which would seem to point to an OE **fæled*, a variant of *falod*, 'fold.' The application of such a name must remain uncertain in view of our ignorance of the topography.

HIND LAKE (lost)

Hindelake, Hyndelake 1223 *FF*, 1279 RH
Hyndelac 1224 (c. 1350) Rams
Hind Lake c. 1750 Bodger's *Map of Whittlesea Mere*

OE *hinda-lacu*, 'hinds' stream,' i.e. where they water. *v.* lacu. On Bodger's map of Whittlesey Mere several such 'lakes' are marked—Henson's Lake, Barnsdale Lake, Long Lake, together with this one on the east side.

HOLME WOOD

boscus de Holme 1248 *For*

SWORD POINT (lost)

> *Sweordora* c. 1000 BCS 297[1]
> *Swerord super Witlemærebanc* 1146 *Cott* vii. 3
> *Swerdeshord super Withelesmere* c. 1150 (c. 1350) Rams
> *Swerdesorde* c. 1150 (c. 1350) Rams
> *Swerdeshord* 1279 RH
> *Swere Point, Swere Hord* c. 1750 Bodger's *Map of Whittlesea Mere*
> *Sword Point* 1766 J, 1787 Cary, 1822 Darton
> *Swere Point* 1836 O

The old maps of Whittlesea Mere show on the south side of the mere a broadish peninsula pushing itself up into the lake. At its north-west corner is a point of land called *Swere Point* and behind it a corner of the peninsula is marked off as *Swere Hord*. Mr Goodall in a study of the document known as the Tribal Hidage (BCS 297), shortly to be published and very generously placed at our disposal, shows how this is to be identified with the *Swerdora* of that document, where it appears as the name of a territory, estimated to contain 300 families, and with the *Swerord* of the foundation charter of Sawtry Abbey (1146). The full significance of this discovery is explained in the Introduction (xix). With regard to the site we can only regret that the draining of the mere has removed all trace of the peninsula and the point. It was approximately where Engine Farm now stands.

The etymology of the name is not an easy one owing to the inconsistency of the early forms. The final element is, almost certainly, ord. That is the form found in all the early documents except the Tribal Hidage, and the text of this document is notoriously corrupt. (The *hord-* forms show the common inorganic *h* which develops before a second element beginning with a vowel.) Further, Bodger's map shows a series of names ending in *-hord* right round the mere, evidently taking their name from projections of land, e.g. *Grimeshord, Alderhord*. What the first element is, is not so clear. *sweord*, 'sword,' whether in the nominative or the genitive, giving a name 'sword(s)-point,' is not very likely on the topographical side, and

---

[1] *Sweodora* in two of the texts.

is inconsistent at the one end with the form in the Sawtry charter which, as a rule, has good forms, and at the other it is difficult to see how it could have developed to the *Swere Hord* found on Bodger's map or could have given rise to a *Swere Point*. On the other hand, if we start with *Swerord*, the compiler of the Tribal Hidage having written *Sweordora* in error for *Sweororda* (with his usual genitival -*a* suffixed to *Swerord*), the development to *Swere Hord* is normal and one can well understand a translation of Swere *Ord* having given rise to a Swere *Point*, while *Sword Point* would be a natural reduction of *Swere Ord Point*. This involves the assumption that the forms in the Ramsey Cartulary and in *Rotuli Hundredorum* are corrupt, but that seems the smaller difficulty of the two, and it may be that a process of folk-etymology was assisted by some such process as a doubling of the suffix, *Swerord* being expanded to *Swerordesord* or *Swerdesorde*. If this is accepted then the *Swerord* is to be explained as a compound of OE sweora, 'neck,' and ord, the whole name meaning 'point at the end of the neck of land.'

Swerord or rather the district to which it came to be applied may have left one other trace on old maps where *King's Delph* is alternatively known as *Sword* Dike or *Swerdes* Delf.

WHITTLESEY MERE

Witlesmere 963–84 (c. 1200) BCS 1128, 1146–53 (c. 1350)
Rams, 1270 *Ass*
Witelesmere a. 1022 (c. 1200) KCD 733, c. 1150 (c. 1300–25)
Thorney 168 *b*
Witelesmare 1086 DB
Witlesmare 1145 *Cott* vii. 3, 1189 (1332) Ch, 1223 FF, 1248–
56 BM
Witthelesmere 1146–53 (c. 1350) Rams, 1270 *Ass*
Witlemare 1146–53 (c. 1350) Rams
Wittesmara 1189 (1332) Ch
Wytlesmer 1260 *Ass*
Wittelysmere c. 1480 BM
Whittelsmere 1535 VE
Wittlesmere 1585 FF

'Witel's mere' *v.* mere. This pers. name is on record (Redin 139) and is a diminutive of the rare *Wita*. The same name is also found in Whittlesey just over the Cambridge border and there is no doubt that *Witles-mere* and *Witles-ig* were named from the same man. The modern form of *Witlesig* has led to a refashioning of *Whittlesmere* as *Whittlesey Mere*. The *Wh-* forms are entirely modern and do not justify Skeat's derivation from an otherwise unknown *Whitel*.

Morborne

MORBORNE   74 B 10

> *Morburn(e)*   1086 DB, 1276 RH, 1286 *Ass*, 1385 Cl, 1428 FF
> *Morborne*   1255 *For*, 1535 VE
> *Morreborne, Molburne*   1286 *Ass*
> *Morbourne*   1316 FA, 1327 *SR*
> *Marborn*   1610 Speed
> *Marborn* al. *Marbon*   1675 Ogilby
> *Marbon* or *Morbourne*   c. 1750 Bowen
> 'Swampy-land stream' *v.* **mor, burna**.

OGERSTON (lost)

> *Ogerestan(e)*   1185 (c. 1200) *Templars*, 1227 Ch, 1253 Pat, 13th AD iv
> *Oggereston*   1189 ChR
> *Oggerston*   1305 Cl, 1360 FF
> *Oggeston*   1335 Orig
> *Ogerston*   1597 FF

The site of Ogerston was marked as Ogerston Ruins as late as the first O.S. map. It was just south of the spot where the Billing Brook crosses the Bullock Road and was on the border of both the parish and county, so that the stan is probably a boundary-stone. The first part is clearly a pers. name, but one cannot accept Skeat's *Ocg-here*. There was an OE name *Ocga*, but one can only explain that as a pet-form for such a pers. name as *Ord-gār*, and it could not in its turn be used as the first element in a compound such as *Ocg-here*. On the other hand there is good evidence for a late OE and EME pers. name *Oger(i)us, Odgar, Odgerus, Ogger* which, as Forssner (197) shows, is an OGer name *Autger, Odger, Og(g)er* which may well have come in through French influence. Derivation from ON *Auðgeirr* or *Oddgeirr* is also possible. *Ogier* le Danois, a medieval hero of romance, corresponds to a Danish *Otgerus*, *Udgerus*. Hence 'Ogger's stone,' the pers. name being one of comparatively late foreign origin.

Water Newton

WATER NEWTON 64 J 9

Niwantune 937 (c. 1300–25) *Thorney 5 b*, 973 (c. 1300–25) ib. 2 *a*, c. 1000 *Cragg*
Newetone 1086 DB
Neuton c. 1150 (c. 1300–25) *Thorney* 169 *b et passim*
Newenton 1291 Tax, 1428 FA
Waterneuton c. 1300 *Linc*
Watter Newton c. 1660 Moore

Self-explanatory. It lies by the Nen.

BILLING BROOK

Bilingbroc 1300 *Ct*
Billyngbroke 1301 *Ct*
Byllingesbroke 1306 *Ct*

Owing to the paucity of early forms it is difficult to say whether this is from OE *Billingbroc* or *Billingabroc* or *Billingesbroc* when we should render it 'Billa's stream' or 'stream of Billa's people,' or 'stream of Billing' or 'of Billa's son.'

Orton Longueville and Orton Waterville

ORTON 64 J 11

ofertuninga gemære 955 (c. 1200) BCS 909
æt Ofertune 958 (c. 1200) BCS 1043
Ovretune, Ovretone 1086 DB
Vuertun' 1158 P
Ouerton 1200 FF
Overton Henrici de Longa Villa 1220 Fees 334
Overton Lúngheuille 1227 *Ass*
Ouertone Wateruile 1248 FF

The first element appears as *Orton* from 1546 (FF). *Longueville* appears as follows:

Longeuill 1286 *Ass*, 1303 FA, 1350–8 Ipm
Lung(e)ville 1296 Cl, 1298 FF, 1314 FF, 1316 FA, 1403 AD ii

Longevyle 1428 FA, 1490 Ipm, 1546 FF
Lungevile 1428 FA
Longvile 1480 AD i, 1492 Ipm
Longfeld 1526 LS, 1561 FF
Langffeld 1527 BM
Longfield 1532 BM
Orton Long al. *Overton Longville* 1641 HMC

Waterville shows, apart from the normal form,

Wautreville 1260 *Ass*
Wautervile c. 1300 BM
Waltervill 1302 Ch
Waterfeld 1548 FF

and is alternatively known as

Chery Orton 1548 FF
Chyrry Orton 1552 FF
Overton Watervile al. *Cherihorton* 1573 FF

a name preserved in *Cherry Orton Farm.*

'Farm by the ofer or bank' (cf. Orton (Nth)). *Longueville* and *Waltreville* are the names of the feudal tenants of the manors, *Cherry*, here, as in Cherry Hinton (C), is a distinctive epithet applied at a late date from the cultivation of cherry-trees. Confusion between *-ville* and *-feld*, pronounced with voiced *f* as *-veld*, is common in p.n. (cf. PN Bk xxv). There are too many French places called Longueville for us to identify the home of this feudal tenant and no source for Waterville has been found.

BOTOLPH BRIDGE

Botuluesbrige 1086 DB, 1224 FF, 1242 Fees 425, 1359 Ipm
Botelesbrig' 1200 FF
Botelbrig(ge) 1220 FF, 1290 Cl, 1359, 1366 Fine, 1369 Cl, 1428 FA
Botulvesbrug 1220 Fees 334, 1366 Cl
Botolfbrigg 1225 FF, 1260 *Ass*
Botolvesbrug 1227 *Ass*
Botilbrig(ge), *Botylbryge* 1286 *Ass*, 1292 Ipm, 1357 Cl, 1358 Ipm, 1428 FA, 1492 Ipm

Botulfbrig 1286 FF
Botylbrugg 1310 FF
Botlebrigge 1327 Ch
Botulfbruge 1348 Ch
Bottelbrigge 1480 IpmR
Bottelbridge 1545 HMC Var iv
Bottle Bridge 1766 J

'Botolph's Bridge,' *Botolph* being the common Norman form
for OE *Bōtwulf*. *Bottle* is the common colloquial development
of it. *v.* Addenda.

GOLDIFORD (lost)

Goldeg 955 (c. 1200) BCS 909
Goldiford 1766 J, 1809 Carey

The ford was by a small island in the Nen called 'Gold-
island,' *v.* eg. The exact significance of the first element in the
name it is difficult to be sure about. It may very well be OE
golde applied to the marigold.

Sawtry, All Saints and St Andrew, and St Judith

SAWTRY 74 D 11

Saltrede 1086 DB
Saltreia, Saltre(y) 1146–53 (c. 1350) Rams, 1157 BM, 1163 P,
    1167 P (*Monacorum*) *et passim* to 1363 Cl
Saltereia, Saltereye 1147 BM, 1152–67 AC, 1186 BM, 1220
    Fees 324 (*Roberti de Bello Mesag'*), 1242 Fees 923 (*Robt de
    Bello Mesnagio*)
Saltreda 1183 P
Sauteria, Sautereye 1184 BM, 1224 Bract, 1228 Cl
Sautre(ye) 1235 Cl, 1258 FF, 1260 *Ass*, 1269 FF *et passim*
Magna Saltreya Monachorum, Majorem Sautr' 1279 RH
Saltreya le Moynne 1279 RH
Sautr' Beumeys 1279 RH
Saltretha c. 1350 Rams
Sawetre 1416 FF
Sawtre Moyns and *Ivet* 1568 FF
Sawtrye Beames 1572 FF

Sawter 1576 Saxton
Sawtry Jewel 1589 FF

There can be little doubt that the second element in this name is OE rið and that the reference is to the stream which makes its way from here down to the fens. *rethe* rather than *rithe* is the form which this word takes in the ME forms of undoubted OE names in *rīð*, as in Shepreth (C), Hendred and Childrey (Berks), Shottery (Wa) and probably also Meldreth (C), the development of *rithe* to *rethe* being due to lack of stress. Loss of intervocalic *th* is very common, as witnessed in two of these very names. The first element may be the simple word *salt* and the term be descriptive of a stream of brackish taste, but it is also possible that, as suggested by the *Saltereia* forms, the first element is OE *sealtere*, 'salter,' and that the full early form of the name was *sealtera-rīð*, 'salters' stream.' If that is the case we should take this as a further example of the numerous names in which *Salt* and *Salter* are found in England, referring to the carriage of salt, that all-important commodity of the Middle Ages. The stream may have been one by which salt was brought up from the fenland water-ways. It may be noted that in the Northants Assize Roll of 1202 there is a reference to tolls exacted for loads of salt passing through Winwick, which may well have come from Sawtry.

Monacorum from the holding of Ramsey Abbey or of Sawtry Abbey itself. *Beumeys* or *Beams* from the family of the Robertus de *Bello Mes(n)agio* who had a holding here in 1220. They had been here since the 11th cent. for we have a Walter de Belmeis here c. 1090 (Rams 229)[1]. Sawtry Monachorum was also known as *Magna* or *Major* Sawtry. Sawtry *Moyne* is named not from the monks but from the Moigne family who were holding here in 1279 (RH). *All Saints* and *St Andrew* are the saints to whom the two churches are dedicated. The Countess Judith, wife of Earl Waltheof, who held land in Sawtry (DB) has been canonised by popular usage in order to keep them company.

AVERSLEY WOOD

Ailbrittesle 1209 *For*

---

[1] The note in PN NbDu (14) should be cancelled in the light of this fact.

Aylbritesle 1245 *For*
Albrichelee (*wood of*) 1290 Misc

'Aeþelbeorht's clearing' *v.* leah, with development of inter-vocalic *b* to *v* as in Pavenham *supra* 36. Cf. also Speed's spelling of Abberwick (Nb) as *Averwick* in 1610 and *Averton* for Abberton (Ess).

MONKS' WOOD

> *le couert de Sautre apele Monkeswod a Moynes de Sautre* 1219 *For*
> *boscus monachorum de Sautr'* 1230 Cl
> *Monekeswode* 1279 RH

Self-explanatory.

Sibson cum Stibbington

SIBSON 64 J 9

> *Sibestune* 1086 DB, c. 1150 (c. 1300–25) *Thorney*
> *Sibeton, Sebeton* 1217 Pat
> *Shipeston* 1218 FF
> *Sibston* 1233 FF, 1324 FF (*juxta Walmesford*)
> *Sibestone* 1279 RH, 1316 FA, 1326 FF
> *Sibiston* 1287 *Ass*, 1327 FF, 1329 BM
> *Sibbiston* 1287 *Ass*, 1329 BM
> *Sebston* 1428 FA
> *Sybton* 1515 FF
> *Sybson* 1544 FF
> *Sybston* al. *Sybson* 1609 BM

'Farm of Sib(b)i,' *Sib(b)i* being a pet-form for such a pers. name as *Sigebeald* or *Sigebeorht. v.* tun.

STIBBINGTON

> *Stebintune, Stebintone* 1086 DB
> *Stibinctuna* c. 1150 (c. 1300–25) *Thorney passim*
> *Stebentun* 1217 Pat
> *Stibeton, Styb-* 1218 FF, 1260 *Ass*
> *Stibenton* 1233 FF
> *Stybinton* 1260 *Ass*
> *Stibington, Styb-* 1291 NI, 1315 Cl, 1316 FA, 1329 BM, 1428 FA

Q

Stepington 1535 VE
Stipingtone 1549 BM
Stybbyngton 1609 BM

This is best explained as from OE *Stybbingtun*, 'Stybba's farm,' **Stybba* being a pers. name allied to OE *Stūf* and *Stȳfic*, and forming the basis of Stebbing (Ess) and Stepney (Mx). Cf. Stevington *supra* 46 and *v.* Zachrisson, *Some English PN Etymologies*, who quotes (p. 132) the form *Stybban snade* from BCS 1054.

ARNEWAS (lost)

Arnewassebrok 1278 Selden ii
Arnewas 1279 RH, 13th AD iv
Arnewessebroke 1348 AD iv

'Earna's water-washed land' *v.* (ge)wæsc. *Earna* is not on record but would be a regular shortened form of one of the OE names in *Earn-*; cf. *Erne, Earne* in DB (TRE).

WANSFORD 64 J 8

Wylmesford 972–92 (c. 1200) BCS 1130
Walmesford 1184 (15th) Gilb, 1224 Bract, 1227 *Ass*, 1244 (c. 1350) Rams, 1279 RH, 1297 Ch, 13th AD iv, 1305 Orig, 1324, 1348 FF
Wammeford 1218 FF
Wanesford 1346 FA (iv. 448)
Walmesford al. *Waynsford* 1589 FF

As BCS 1130 is a record of land acquired for Peterborough Abbey by its second founder, Æthelwold, Bishop of Winchester, it may very well have been written by a West Saxon clerk and the earliest and later forms of this name can then be reconciled. *Wylmes* will be the West Saxon form, and *Walmes* will be from the Anglian form *wælm* of the same word. The latter is the one we should normally expect in this area. The meaning would be 'ford of the spring or stream.' As a stream flows into the Nen just opposite Wansford village, the latter meaning seems entirely appropriate.

WANSFORD BRIDGE (6″)

pontem de Walmesford 1286 For

Stanground

STANGROUND 64 J 12

Stangrund c. 1000 *Cragg*, c. 1150 (c. 1300–25) *Thorney* 168 *b*,
1270 *Ass*, 1285 (c. 1350) Rams, 1301 BM, 1302 Ch, 1307
FF, 1316 FA
Stangrun 1086 DB
Standgrund 1276 RH
Staingrunt 1286 QW
Stainground 1286 *Ass*
Stanground 1327 Cl, *SR* (p)
Staneground 1330 FF
Stanegrund 1428 FA
Standground 1641 HMC

This is clearly a compound of OE stan and grund and must
refer to the stony character of the ground or of the bottom of
some water, but its exact application cannot be determined.

HORSEY HILL

Horesheya 1219 *FF*

The forms are insufficient for certainty, but topographically
the explanation 'horse-island' would suit well for Horsey Hill
is a slight fort-crowned elevation amid the marshes.

Stilton

STILTON 74 C 11

Stic(h)iltone 1086 DB
Stichelton 1167 P
Stikelton 1181 P (p)
Stilton 1219 FF *et passim*
Stileton 1227 *Ass*, 1358 FF
Stig(h)elton 1227 *Ass* (p), 1228 Cl[1]

We may take the first element to be OE *stigel*, 'stile,' hence
'farm marked by some distinctive form of stile.'

[1] The forms in the *Lincoln Registers* are *Stilton* (1290–1360), *Styelton*,
Steelton, *Styltone* (c. 1390), *Stielton*, *Stelton* (c. 1410), *Steleton* (c. 1420 and
c. 1560).

Washingley

WASHINGLEY   74 C 10

Wasingelei 1086 DB, 1286 *For*

Wassingelai, -lea, -le(ga) 1163, 1167, 1187 P, 1198 *P*, 1241 FF, 1245 *For*, 1292 Ipm, 1294 Cl

Wassingle(y), -yng- 1185 (c. 1200) *Templars*, 1260 *Ass et passim*

Wassynghele 1260 *Ass*

Wassigleye 1261 FF

Washingle 1286 *Ass*

Wasinglee 1286 Orig

Wasshyngle 1429 IpmR

Wasshelyngle 1518 FF

The *Washing-* names in English place-names are not easy to explain.   The essential material is found in Washington (Sx), *Wasingatun, Wassingatun, Wessingatun* in BCS 1125, 834, 819, all 12th cent. copies of Saxon charters, Washingford (Nf) DB *Wasingaford*, later *Wasingford, Wassingford*, Washingborough (L), DB *Washingeburg* and Washington (Du), with forms in *Wess-* and *Wass-* and, after 1300, in *Wessh-* and *Wassh-*, and one in *Quess-* in 1280. Washingborough also occasionally has forms with initial *Qu-*. In addition to these we must also note Washbourne (Gl) which is *Wassanburna* in 11th cent. copies of Saxon charters (BCS 236, 430) and *Waseborne* in DB, and also a lost *Wasincham* in the Norfolk Domesday, and a lost *Watsingaham* (Sr), BCS 693.

In interpreting these forms it should be noted in the first place that the history of Washbourne makes it practically certain that there was an OE pers. name *Wassa* (cf. OGer *Wassingun, Wassenstein*, Förstemann PN) and that this might at a later date appear as *Wash-*, for it is impossible to believe that an OE name with initial *Wæsc-* would appear in this form in the charters in question. The same is true with equal certainty of the old forms of Washington (Sx). We cannot be so sure in the case of the other place-names for which we have only ME forms, for *ss* in ME may represent the *sh-* sound. It should be noted, however, that, with the exception of Washingborough, forms with distinctive *sh* are late in appearance, and one may suspect that all except this originally had a pure *s* or *ss*. If that is the case we should interpret Washingley as 'leah of Wassa's people' and presume that *Wassingley*, like so many of these names, was changed to *Washingley* under the influence of the common word *wash*. On the other hand, it is just conceivable that here and in one or two other of the *Wash-* names, we have, as the first element, *Wæscingas*, 'dwellers on the *wæsc* or small stream,' and that the whole name means 'clearing of the dwellers on the wæsc.' In some cases there are clearly such streams and we need not look for anything like the Wash itself[1].

CALDECOTE WOOD

boscus de Caldecote 1248 *For*

v. ceald, cot.

WARD MOUND (6″)

Wardhowe 1241 *FF*

If this identification is correct the meaning would seem to be 'watch-hill' (*v.* weard, hoh). It is on high ground, though not quite on the highest ground in the locality. There is an interesting earthwork site, the date and origin of which are uncertain[2].

Yaxley

YAXLEY 74 A 11

æt Geakeslea 955–9 (c. 1300–25) *Thorney* 6 *b*

Geaceslea 963–84 (c. 1200) BCS 1128, c. 1000 *Cragg*

Iaceslea c. 970 (c. 1200) BCS 1131

Geakeslea, opergeakeslea 973 (c. 1300–25) *Thorney* 1 *b*, 3 *b*

[1] Since the above explanation was written Professor Zachrisson has very kindly sent us an advance-proof of his forthcoming *English PN and River-names containing the primitive Germanic roots Vis, Vask*. It reached us as copy was being passed for press and it was impossible to give his most valuable paper the consideration that it calls for. Suffice it to say here that it is all in favour of taking the names here set forth as containing OE *wæsc* or (possibly) *wāse*, 'mud,' rather than as containing a personal name. In support of his view, in at least one case, we may note that the Surrey *Watsingaham*, as noted by him, certainly comes, not from an original charter but from a 12th cent. text and the *ts* is an AN spelling. Further, Mr C. S. Seyler in a minute study of this charter, kindly placed at our disposal, has shown that this lost manor may have stood on a small stream, whose course he has traced, which left its name in 'The *Wash* Way,' near the site of the present Clapham Junction Station.

[2] *ex inf.* Mr C. C. Tebbutt.

Iacheslei 1086 DB
Jakeslea Wm i (c. 1300–25) *Thorney* 79, c. 1150 ib. 168 *b*
Iakesle(e) 1203 FF, 1227 BM, 1232 Cl, 1260 *Ass*, 1342 Ch, 1347 Ipm
Jackele 1227 *Ass*
Jakesle(ya) 1227 *Ass*, 1275, 1322, 1379 Cl
Jakle 1227 Ch
Iakele(ye) 1253 FF, 1255 *For*, 1276 BM, 1286 *Ass*, 1302 Ch, 1327 *SR*
Iaskele 1297 FF
Yakesle 1302 FF, 1327 Cl, 1345 FF, 1364 Cl
Yaxlee 1389 AD vi
Yakesley al. *Yaxley* 1591 FF

OE *gēaces-lēage* (dat.), 'cuckoo's clearing,' so called because infested by such. *v.* leah.

DRAYMERE (lost)[1]

Dreigmære a. 1022 (c. 1200) KCD 733
Draymere 1279 RH, 1286 *Ass*
Dray Meer 1766 J
Dray Mere 1836 O

This lost mere gives us another compound of the puzzling word dræg discussed in EPN. Possibly the reference here is to the use of a *dræge*, i.e. a *dray* or drag-net for fishing in the mere, or the reference may be to some cut through the Fens.

TRUNDLE MERE

Trendelmere 955–9 (c. 1300–25) *Thorney* 7 *a*
Trendmære a. 1022 (c. 1200) KCD 733
Trendelmare 1279 RH
Trandley mire 1516 Saxton

This compound describes the shape of the mere, the first element being OE *trendel*, 'ring, circle.'

WALLPOOL (lost)

Wellepol a. 1022 (c. 1200) KCD 733
Wallpool Pit 1766 J, 1787 Cary

---

[1] On Jefferys' map it is marked just to the south of Trundle Mere.

'Pool with or by the stream or spring' v. **wielle, pol.** The variant vowel forms are due to the same cause as that noted under Wansford *supra* 198.

YAXLEY FEN

Mariscus de Jakele 1286 *Ass*

YAXLEY LODE

Jackeslada 1227 Ch

v. (ge)lad. The common term for water-ways in the Fen district.

II. HURSTINGSTONE HUNDRED

Hyrstingestan, Hertingestan 1086 DB
Herstingestan 1086 DB, c. 1120–30, c. 1136–40 BM
Hurstingestan 1168 P
Horstingestan 1169 P
Hurstincton 1189 BM
Hirstingestan 1207 P, 1270 *Ass*
Hirstlingestan 1209 For, 1227 *Ass*
Hurstingeston 1227 *Ass*
Hyrstingston, Hirstyngston 1255 *For*, 1327 *SR*, 1428 FA
Hurstyngston, Hurstingston 1303 FA, 1585 D
Hirstlyngstone 1364, 1370 Cl

The history of this Hundred-name has been made out by Mr Goodall in the paper already referred to *s.n.* Sword Point *supra* 190. He makes a convincing case for thinking that under the immediately preceding tribal name, the *Herefinna* with 1200 families, are concealed the *Hyrstingas*, who gave their name to this Hundred. This is justified, not only by their position, but also by the evidence of the MSS themselves, for as Mr Goodall points out, the Latin version of this document (BCS 297 A) has *Herfuina* with an alteration in a late copy to *Herstina*, while version 297 B has *Heresinna*. These *Hyrstingas* must have been so called because they lived in the wooded district which included Old and Wood Hurst, Upwood, Wood Walton and Warboys, places which, by their very names, remind us that this district was once well wooded. The meeting-place of these *Hyrstingas* or woodland dwellers was at the stone which is still marked as the *Hursting Stone* by Gordon and Bowen in their 18th cent. maps of Hunts. It now appears as the Abbot's Chair, a mile and a half to the south of Old Hurst. The Hurstingstone Hundred was held by Ramsey Abbey and it is clear that from the 12th cent. the Abbot held his court at the old meeting-place. The Chair is a large square stone, in shape of a chair, with traces of an inscription, now illegible.... Local tradition says that Mother Shipton used to sit in this chair and utter her prophecies in the 16th cent. The stone must have come from a distance, as there is no stone in this part of Hunts[1]. *v.* Addenda.

Bluntisham cum Earith

BLUNTISHAM [blʌntsəm], [blʌntʃəm] 75 G 4

  Bluntesham Edw Conf (12th) KCD 907, 1086 DB, 1103–31 BM *et passim*
  Bluntsome 1545 BM
  Blunesham 1549 Pat
  Bluntsam E 6 BM
  Blunsham Eliz ChancP

'Blunt's homestead' *v.* ham. It is clear that there must have been a pers. name *Blunt* in OE for it is found in Blunsdon (W), DB *Bluntesdone*, *Blunteshale* in Essex (cf. Colchester Cartulary ii. 526), Bluntington (Wo), *Bluntesig* (KCD 666) and in *Bluntes-diche* in Needingworth (Ramsey Cartulary 294). In this last case, however, the reference is probably to the same person as in Bluntisham itself. The name is doubtless the same as the ordinary word *blunt*, but its origin is as obscure as that of that word.

EARITH

  Herhethe 1244 (c. 1350) Rams
  Earheth 1260 *Ass*
  Erehithe 1260 *Ass* (p)
  Erheth 1279 RH, 1286 *Ass*
  Ereheth 1286 *Ass*, 1318 Ch

[1] *ex inf.* Mr R. C. Gardner.

Herhyth 1350 AD i
Erethe 1548 BM
Eryth(e) 1557 FF, 1616 BM

We must have here and in Erith (K), as suggested by Skeat (PN C 34), a compound of OE *ēar* and hyð, for the OE form of the Kentish name is *Earhyð* (BCS 87). If so the meaning is perhaps 'muddy landing place.' *ēar* is only found as the name of one of the runic letters in OE, and in the *Runic Poem* it probably denotes 'earth.' The ON cognate *aurr* is used of wet clay or loam, and as these places are 'hithes' we are probably right in taking the sense of the OE word also to be 'mud' here. The situation of the place, immediately upon the bank of the Ouse, suits this derivation.

EARITH BRIDGE

  pontem de Herithe 1219 *For*

GULL FIELD (6″)

  Ye Gulls 1712 *Terr*[1]

As the earlier form shows, this has nothing to do with the bird but contains the word *gull*, apparently a variant of *gool* (OFr *goule*), still used dialectally of a channel made by a stream and found in Goole (Y). It is related to the words *gully* and *gullet*. The field is intersected by water-courses.

HERMITAGE (local)

  Hermitage is an important spot in Earith where there was an old passage over the Ouse. There is evidence of a hermitage here in the 14th cent. Locally it is called *Armitage*[1].

WEST FEN (6″)

  Westfen 1253 *For*

WHITE BRIDGE

  album pontem 1294 *Ct* (BM)

These three names are self-explanatory.

----

[1] *ex inf.* the Rev. Edward Peake and Mr H. E. Norris.

Broughton

BROUGHTON[1] 75 F 1

Broctune 1086 DB, n.d. (1300–25) *Thorney* 3 *b*
Brocton 1202–7, 1245 BM
Broucton 1254–67 BM, 13th AD ii
Brochton 1260 *Ass*
Broughton 1303 FA *et passim*
Broghton 1305 FF
Browtun 1546 BM

'Brook farm' *v.* broc, tun.

WOOD (Fm and Lane) (6″)

cf. *Wodedole* 1300 *Ct*

Self-explanatory.

Bury

BURY 75 D 1

(æt) *Byryg* c. 1000 *Cragg*
Biria, Birig, Byrig, Biri, Biry, Byri 1100–35 (c. 1350) Rams,
   1253, 1297 BM, 1300 *Ct*, 1311 AD v, 1327 *SR*, 1334 BM,
   1336 AD i
Bury 1359 FF
Bery 1404 AD i
Bury al. *Berry* 1564 FF
Bery Parua 1569 FF

The *burh* referred to in this name is probably the earthwork described as the 'Roman Camp' just to the south-west of the village.

CHEVERIL LANE (6″)

cf. *Cheverheth* 1306 AD i

---

[1] Through the kindness of Dr R. H. Murray, Rector of Broughton, the following survivals of early field-names have been noted:

Ramsey Cartulary (1252) *Waterwolde, Holdedole, Stanilande, Buttes ad Rowdych, Grenehyll* as Oldwaters, Hold-dole, Stonelands, Buttes-close, Grennills.

Court Rolls (1293) *Brokenhyl, Saltwell, Haycroft* as Brokenback, Salt-wells, Haycraves.

Ancient Deeds (13th cent.) *Hullok, Rypthornes* as Hullucks, Ripthornes.

HEPMANGROVE (lost)[1]

Height-, Heyt-, Heyghtmond(e)groue 1297 *Ct*, 1377 FF, 1378
   Cl, 1437 AD i
Heyt(h)mundegrave 13th AD i, 1303 AD vi, 1309 AD i, 1311
   AD v
Eydmundgrave c. 1300 BM
Heymundegrave 1327 *SR*
Heyghnundegrave 1338 AD i
Heighmondegrove 1359 FF
Hetmingrove 1387 IpmR
Hetmigrove 1392 BM
Hetmun(d)grove 1404 AD i
Hecmegrove 1428 AD i
Heghtmongrove 1484 FF
Hekmangrave 1498 BM
Heyghmongrove 1516, 1540 BM
Highmongrove 1517 FF
Hethemongrove 1542 BM
Hepmangrove 1552 FF
Highmondgrove 1558 BM
Hemyngrove 1569 FF, 1570 BM

'Hēahmund or Hæðmund grove,' with a curious absence of
any sign of the genitival form of the pers. name (*v.* graf(a)). In
some ways an OE *Hēahmund*, which is actually on record (while
Hǣðmund as a compound has not survived and in itself is less
probable), would fit the early forms better, especially so far as
they show diphthongal *ey, ei*. If we start from that we should
have to explain the *t* and *d* forms as AN spellings due to the
difficulty which AN speakers and writers found in dealing with
OE *h*. The variation between *t, k*, and *p* is curious but has its
parallel, though not in the same order of development in the
history of Pavenham *supra* 36. The same pers. name is found
in Heathencote (Nth) for which we have forms *Heymondcot*
(1284 FA), *Heymundecotes* (1308 Ch), *Heitmundecotes* (1337
Ipm), *Hegmondcot* (1428 FA). The phonological development
of this name is equally difficult.

---

[1] On Bowen's map (c. 1750) the site is just west of Bury.

NORTHEY (Fm) (6″)

Northeya c. 1350 Rams

As it is in Bury Fen the meaning is clearly 'north island,' *v.*
eg. It lies on the northern boundary of the parish.

Colne

COLNE 75 G 4

Colne Edw Conf (12th) KCD 907, 1086 DB *et passim*
Colen, Collen 1279 RH
Cone c. 1660 Moore, *Map of the Great Levell*

This would seem to be an old stream-name, identical with
that of the river Colne (Ess, Mx).

Hartford

HARTFORD 74 H 14

Hereforde 1086 DB, 1199 Cur, 1223 FF, 1260 *Ass*, 1276 RH,
    1285 Cl, 1286 *For*, 1308 Fine, 1327 Cl, 1495 BM, 1535 VE
Herford 1147 BM, 1194 Cur, 13th AD i, 1444 IpmR, c. 1350
    Linc (*juxta Huntingdon*)
Harford 1410 BM, 1535 VE, 1558 FF, Eliz ChancP
Hertford 1428 FA, 1526 LS, 1529 FF, 1542, 1558 BM

'Army ford' *v.* here, ford. The *t* is intrusive and late.

SAPLEY

Sappele 1227 ClR, 1232, 1235, 1236, 1238, 1245 Cl, 1255 *For*,
    1260 *Ass*, 1299 BM, 1300 Rams, 1367, 1381 Cl
Sapele 1232, 1235, 1240, 1243 Cl, 1255 *For*
Shappele 1238 Cl
Sap(p)le 1245, 1378 Cl
Sappeleye 1275, 1285 Cl
Sappele Hey 1292 Cl
Sapperleye 1338 Orig

OE *sæppe-leage* (dat.), 'fir-tree clearing,' *v.* sæppe, leah. If
this is correct it is good evidence, in this old forest district, for
the presence of firs in English woodland at a good deal earlier
date than is usually allowed.

Holywell cum Needingworth

HOLYWELL [hɔliwəl] 75 H 3

Haliewelle 1086 DB
Haliwell, Haly- 1231, 1234 FF, 1238 Cl *et passim* to 1526 LS
Halliwell, Hally- 1350 AD i, 1571 FF, 1601 BM
Haylywell 1569 FF
Hollywell 1600 FF, c. 1750 Bowen
Hallowell Eliz ChancP, 1601 Cai

'Holy spring' (*v.* halig, wielle) with the common shortening of the vowel in later *Halli-* and *Holli-*. The modern spelling is due to the influence of the uncompounded adj. *holy*. The Holy Well itself still survives and is known as such[1].

MOYNES HALL

Moynes Hall Eliz ChancP

It seems very probable that this is the messuage with garden and 1 carucate of land and 10 acres of meadow which the Abbot of Ramsey held in Needingworth of the gift of Sir Berenger *le Moyne* in 1279, RH ii. 602 (*VCH*).

NEEDINGWORTH [ni·dinə·θ]

Neddingewurda 1161, 1163 P
Nithingwurth 1227 *Ass* (p)
Nedingewrht 1234 FF
Nidingw(u)rth 1241 FF, 1260 *Ass*, 1268 FF
Niddingworth 1260 *Ass*, 1287 *Ct*, 1317 AD i, 1327 *SR*, 1342 FF, c. 1350 Rams
Niddingeworth 13th AD iv
Nyd-, Nidingworth 1322 AD i, 1337, 1417 FF

---

[1] Through the kindness of the Rector of Holywell (the Rev. J. A. Ross), we have received a list of field and other names surviving in 16th and 17th cent. deeds, churchwardens' accounts, etc. The most interesting of these are *Lownde Field* and *Lownde Hole* from a deed of 1619, and still surviving in *Lowndes*. This is clearly the *Hunderlunde* of the Ramsey Cartulary (i. 294) and another example of ON *lund* (cf. Toseland and Holland). In confirmation of this Mr Ross draws our attention to the fact that in a 17th cent. document we have reference to 'the grove of woods called ye Lownd.' As Lowndes lies at the foot of a slight hill Mr Ross suggests that the mysterious *Hunder* is for *Under-*. In addition we may note *Lowdells* or *Lowdelves*, a field-name in Needingworth, showing the same variation in forms that we get in *Shelve* or *Shell* (Wo).

Nedyngworth, -ing- 1452 FF, 1456 AD i, 1535 VE
Needenworth 1662 Fuller, 1675 Ogilby

Ekwall has provided the solution of this name when (PN in *-ing* 14) he associates it with Nedging (Sf). The latter is found as *Hnyddinge* (BCS 1289), and he takes this to be from a lost OE name *Hnydda*, probably by origin a nickname, allied to English *nod*, *noddle* and *nuddle* (dial.), 'to push.' Needingworth is then 'enclosure of Hnydda's people' (*v.* worð). Short *y* appears quite regularly as short *e* or *i* in ME (*v.* Introd. xxv). The lengthening of the vowel, first clearly shown in the 17th cent., may be purely artificial and due to association with the ordinary word *need*. If it is a regular development then we must take it that the double consonant *dd* early came to be regarded as a single *d* and that *i* was lengthened, lowered to *e*, and made tense in the open syllable, but this seems somewhat unlikely. It is interesting to note that the only other known example of this pers. name in place-names is a lost *Nidingham* (C), so that all the examples of it come from the eastern part of the region settled by Angles.

STOCK'S BRIDGE (6″)

Stockes (cultura) 1254–67 AD i

If this identification is right then the proper form of the name is 'Stocks Bridge' and it means 'bridge by the stumps.' *v.* stocc.

Houghton

HOUGHTON 75 H 1

Hoctune 1086 DB
Hocton(e) 1207 P, 1227 *Ass*
Hou(c)hton 1240 FF, 1303 FA
Hohtun 1253 BM
Houton 1279 RH (*cum Wytton*), 1295 FF, 13 AD i (*by St Ives*)
Houghton 1309 BM

'Farm at the foot of the hoh,' the *hoh* being the hill spur otherwise known as Houghton Hill.

HOUGHTON HILL

atte Hyl 1252 (c. 1350) Rams (p)
Self-explanatory.

Old Hurst

OLD HURST 75 F 2

Waldhirst 1227 *Ass*
Waldhurst 1227 *Ass*, 1228 FF
Hirst 1228 FF, 1285 BM
Wald(e)hyrst 1252 BM, 1272 FF
Woldhirst, -hyrst 1258 FF, 1272 (c. 1350) Rams, 1272 FF,
    1294 AD i, 1318 FF
Weldhurst, Weldhirst 1270 *Ass*, 1355 BM
Woldhurst 1350 FF, 1546 BM

The name *hyrst* must once have been applied to the whole
district round here, from its well-wooded character (see further
Hurstingstone *supra* 203).

In course of time when some of the woodland was cleared
the more open country round Old Hurst was distinguished as
Wold Hurst (*v.* **weald**), while the still thickly wooded country
round *Wood* Hurst came to be distinctively so called. Curiously
enough, at the present time there are small woods at Old Hurst
but none at Woodhurst. The loss of initial *w* is fairly common
dialectally (cf. Odell *supra* 34 and Old (Nth)).

Pidley cum Fenton

FENTON 75 F 2

Fentun 1236 FF
Fenton 1279 RH

'Fen-farm' *v.* **fenn, tun.**

FENTON LODE

Fentonelode 1286 *For*
lada de Fentone 1294 (c. 1350) Rams

One of the fen-land lodes or water-ways.

PIDLEY

Pydele, Pid- 1228 Ch, 1260 *Ass*, 1279 RH, 1286 *Ass*, 1387
    IpmR, Cl
Podel 1286 *Ass*
Pudele 1319 Fine, 1327 Cl, Pat
Puddele 1319 Ipm

Pedele(y) 1319 FF, 1526 LS
Pydley 1535 VE

'Pyd(d)a's clearing' *v.* leah. Cf. Piddington (Nth) and Pedley in Clophill *supra* 146.

PIDLEY HILL (6″)

Pydelehille 1252 (c. 1350) Rams

Self-explanatory.

ROWEY (6″)

Rueyemere 1279 RH
Rowey(e) 1286 FF, 1337 Cl, 1387 FF, 1410 AD ii
Rounhey c. 1350 Rams
Rohey 1512 FF
Rowhey al. *Rowey* 1584 FF

Probably OE *rūgan ēge* (dat.), 'rough island,' judging by its situation in the Fens (*v.* ruh, eg). The *h* is the common inorganic *h* found before a second element beginning with a vowel.

STROUD HILL (6″)

Strode 1228 Ch

v. strod. 'Marshy ground overgrown with brushwood' would aptly fit the site.

Ramsey

RAMSEY 75 D 1

Ramesige 1034 C (11th cent.) ASC
Rammesege 1050 D (c. 1050) ASC
Ramesege c. 1060 (12th cent.) KCD 853, 1045 D (c. 1050) ASC
Hramesege 11th BM
Ramesia R i BM
Rames(eie) 1200 FF
Rammes(eye) 1227 *Ass*

After this forms with one and two *m*'s respectively are about equally common.

This is, almost certainly, 'Raven's island,' the first element being an OE pers. name *Hræfn* which must have existed quite independently of the late OE *Rafen* from the ON *Hrafn*.

The existence of this name in OE is made certain by such place-names as Raveningham (Nf), as well as by the numerous place-names in pre-Conquest charters which contain some form or other of this word in place-names outside the Scandinavian area (cf. Ravensden *supra* 61). It is impossible to believe that the reference is to the bird in the majority of these place-names. Such are Ramsbury (W), *Rammesburi* (BCS 828), Ranscombe (Ha), *Hremmescumb* (ib. 938), Romsley (Sa), *Hremesleage* (KCD 1298), Ravenshill (Wo), *Ræfneshyl* (BCS 356), and numerous other unidentified examples. The *h* was early lost and *fn* assimilated to *mm*. For the conditions under which such a name might arise see under Ravensden[1].

ASHBEACH (Fm)

Assebeche 1286 *For*, *Ass* (p), c. 1350 Rams
Osbich 1601 BM

'Ash-beach,' descriptive probably of a farm with ash-trees on the edge or 'beach' of a mere, in the days before the draining of the Fens. Cf. Chalderbeach *supra* 188.

BEAUPRE DROVE (6″)

Beaurepeir 1286 *For*
Beaurepier c. 1350 Rams
Beauraper al. *Bewpre* c. 1500 BM

A common p.n. of French origin, from OFr *beau-repaire*, denoting 'beautiful retreat.' It is specially common in connexion with monastic property and is found in Beurepair (Ha, K), Belper (Db), Bear Park (Du) and Berepper (Co). The pronunciation with initial [bju] is a common development of OFr *beau*, as in Beaulieu (Ha), Bewdley (Wo). An alternative development is that found in Beadlow *supra* 147.

BIGGIN

Bigging, Byggyng 1262 AD i, 1547 FF
le Bygginge 1286 *Ass*
Biggin 1552 FF
v. bigging.

---

[1] One need only record as a curiosity the derivation, found in a life of St Oswald (*Historians of the Church of York*, i. 432), of Ramsey from the Latin *ramis*, 'branches,' for 'the island is as it were hedged round by great trees.'

R

BODSEY HOUSE

Bodes(h)eye 1216–41 (c. 1350) Rams, 1220 FF
Botsey 1610 Speed

There is adequate evidence for an OE name *Boda*, contained in Bodenham (He) and Bodicote (O) (*v.* Redin 45, 121). There is no OE evidence for a strong name *Bode*, and it is probable that the first element of the present is a short form of a compound pers. name ending in *-bod*, such as *Winebod*. 'Bod's island' *v.* eg.

BRICK MERE (6″)

Birchemere 1146 *Cott* vii. 3, c. 1150 (c. 1350) Rams, 1279 RH
Birkemare 1248–57 BM
Berch(e)mere 1217–72 (c. 1350) Rams, 1279 RH
Brichmerpoles c. 1350 Rams
Brikmere 1572 BM, 1610 Speed

OE *beorc-* or *beorca-mere*, 'mere of the birch-trees,' with late metathesis of the *r*. *v.* **beorc**. The presence at one time of birches in the neighbourhood is confirmed by the reference in Dugdale, *History of Embanking* (2nd ed.) 357, to *Byrchholt* by *Byrchmere* (cf. *Bricheholt* in Higney in Ramsey Cartulary).

BROADALL'S DISTRICT etc.

Broadwall Fen 1766 J
Broadalls 1854 Hatfield's *Directory*

The modern form is corrupt. The old name may have meant 'broad spring or stream,' *v.* **wielle**, and cf. Wallpool *supra* 202.

DAINTREE FARM

Dauintre 1260 *Ass* (p)
Daventre c. 1350 Rams (p), 1405 BM
Dauntree 1636 Hayward, *Survey of the Fens*
Dantrey 1695 Morden's *Map*

This must be a late name derived from some man coming from Daventry (Nth), DB *Daventreu*, of which the popular form is *Daintry*. There is a similar Daintree Farm in Cambridgeshire.

GREEN FARM (6″)

le Grene 1363 AD i
Self-explanatory.

THE HERNE
  la Hern 1219 *For*
  Herne 1251 Ch (p)
  þe Hirne 1275 (c. 1350) Rams (p)
  le Hyrne c. 1350 Rams
  'Corner' *v.* hyrne. This is the name given to the area which
forms the extreme north-west corner of Ramsey parish.

HIGNEY (Fm)
  Hyggeneya, Higg- 1146 *Cott* vii. 3, c. 1154 (c. 1350) Rams,
    1235 FF, 1279 RH, 1314 BM
  Higkeneia 1154 (c. 1350) Rams
  Hykeneye, Hik- 1260 *Ass* (p), 1299 BM
  Hygeneye E 1 BM, 1279 RH, 1321 AD i, 1327 *SR*
  'Hycga's island' (*v.* eg), *Hycga* being a regularly formed pet-
form for an OE name in *Hyge-*. Cf. Hughenden in PN Bk 182.

HOOK'S LODE (6")
  Hokeslade, Hokeslode 1279 RH
  'Hoc's water-way' *v.* (ge)lad.

MIDDLE MOOR
  Middelmor 1286 *For*
  Self-explanatory.

MONKS' LODE
  ladam Monachorum 1279 RH
  Munkeslade c. 1350 Rams
  'Monks' water-way,' so called from the monks of Sawtry
rather than from those of Ramsey.

NEW FEN (Fm)
  le newefeldfen 1303 AD v
  Self-explanatory.

RAMSEY MERE
  Ramesmere 13th (c. 1350) Rams, 1335 Pat
  'Mere of *Hræfn*,' the same man who gave his name to Ramsey
itself. Cf. the history of Whittlesey Mere in relation to Whittlesey *supra* 192.

STOCKING FEN (6")

Stokkyngfen 1387 AD ii

'Fen by the clearing' *v.* stocking.

UGG MERE

Ubbemærelade a. 1022 (c. 1200) KCD 733
Ubbemere 1146 *Cott* vii. 3, c. 1150 (c. 1350) Rams
Hubbemare 1192 (c. 1350) Rams
Ubbemare 1248–57 BM
Hubbemerebeche 1252 (c. 1350) Rams
Ubmere 1342 (c. 1350) Rams, 1345 ib.
Ubmere, Ugmere 1572 BM

'Ubba's mere' *v.* mere, some of the forms referring to a 'lode' which connected Ugg Mere with Whittlesey Mere and others to its 'beach' or shore. The modern *g* seems purely irrational, but cf. Pavenham *supra* 36. The OE pers. name *Ubba* is well recorded, and occurs in the p.n. Upton Lovel (W). The corresponding ON *Ubbi* may however be represented here. The pers. name *Ubbe* which was still used in East Anglia in the 12th cent. may represent either the OE or the ON name.

UGMERE COURT ROAD (6")

Ubmerecote c. 1230 (c. 1350) Rams
Hubbemerecote 1279 RH, c. 1350 Rams

'Ugmere cottages,' with the common rendering of *cote* as *court*, *v.* cot.

GREAT WHYTE (6")

la Wihte 13th AD ii
le Whyte 1323 AD i
la Whygthe 1337 AD i
le Wyghte 1349 AD v, 1350 AD i
le Wight, le Lytilwight 1419 AD iii
le magna Wygth 1443 AD i
le Whight magna 1455 AD ii

It is clear that the OE name for this place was *wiht*, and one may perhaps quote *Wi(h)thull* (KCD 709), now Whitehill in Tackley (O), but no suggestion can be at present made as to the interpretation. *v.* Addenda.

WORLICK

Wiluuerihc al. *Wyltherik* 1242 BM
Wolrewok (p) c. 1280 (c. 1350) Rams
Wyrlewyk 1287 *Ct*
Wurlewyk 1382 BM
Worlyche 1536 FF (p)

The early forms of this name are too late, corrupt and divergent for interpretation. *v.* **wic.** *v.* Addenda.

Great and Little Raveley

RAVELEY 74 E 14

Ræflea c. 1060 (c. 1350) Rams, 1077 (17th) Chron Rams
Rauelai 1163 P
Rauelea et altera R. 1167 P
Ravele(ye) 1227 *Ass* (*et passim*)
Rauesle 1228 FF
Magna Rauele, Graunt Rauele 1297, 1364 BM
Raveley Magna and *R. Moynes* 1543 FF

This is a very difficult name. Skeat suggests that there was an OE pers. name *Ræfa*, a shortened form of various names which he quotes. The first of these is *Ræfcytel*, but that is only Searle's faulty reconstruction of DB *Rauechetel* which is clearly ON *Hrafnketell*. Of the others *Ræfwine* is possibly OE, *Ræfmǣr* and *Ræfnōth* are Searle's reconstructions from DB *Rauemerus* and *Rauenod* of uncertain origin, while *Ræfweald* and *Ræfwulf* are reconstructions of English cognates for the OGer *Rafold* and *Rafolf* which are actually on record. There is then some evidence for OE names in *Ræf-* though it is not very strong, and the history of this element both here and in its German cognate is obscure. Derivation from a pers. name is made improbable by the form *Ræflea*, which is confirmed as being correct by its appearance in William i's charter of 1077 (Chron Rams 200) of which Spelman saw the original text. Raveley *Moynes* is Great Raveley, and takes its name from the Moyne family who owned it.

POPLAR SPINNEY (6″)

Popely 1228 *FF*
Popelenge 1228 (c. 1350) Rams

The identification is uncertain though probable. We may compare the equally difficult Popes Wood (Berks) which in an original charter of 949 appears as *wopig hangra* (BCS 877)[1] and in a 12th cent. copy of a charter of 956 as *popping hangra* (BCS 963). This suggests a pers. name *Poppa*, belonging to the obscure group of names of which *Pappa* and *Pippa* are examples.

RAVELEY WOOD (6″)

    boscus de Rauel' 1218 FF, 1248 *For*

REDLAND HILL (Spinney) (6″)

    Redeland 1252 (c. 1350) Rams

Either *hrēod-land*, 'reed land,' or *rēade-land*, 'red land.' Topographical enquiry has not settled the question.

TEN ACRE SPINNEY (6″)

    Tenacres 1252 (c. 1350) Rams

Self-explanatory.

Abbot's and King's Ripton

RIPTON 74 F 13, 75 G 1

    Riptune 1086 DB

    Ripetona c. 1139 BM

    Ripton 1163 P (*Abbatis, Regis*), 1227 Ass, 1231 FF (*Magna*) et passim

    Rupton 1209 *For*

    Ripptune 1253 BM

    Kyngesripton 1381 Cl

    Ripton Abbottes al. *Saynt John's Ripton* 1579 FF

    Rippon 1675 Ogilby

Skeat suggests that this name contains the gen. sg. of an OE pers. name *Rippa* inferred from *Rippanleah* (KCD 1031), now Ripley (Db) and entering into Ripley (Db, Y). The fact, however, that in all the forms of Ripton that have been noted,

---

[1] This is the correct reading of the MS as very kindly collated by Miss F. E. Harmer. It seems almost certainly to be an example of the not in frequent confusion of *w* and *þ* in OE MSS.

with one solitary exception, there is no trace of any vowel between the *p* and the *t* seems definitely to rule out this explanation. More probably we should connect it with Ripe (Sx) for which Roberts (PN Sx) gives forms *Ripe, Ripp, Ryppe, Rype.* Much more important is the fact that it is almost certainly to be identified with the 'silbam qui appellatur Ripp' of an original 8th cent. charter (BCS 160) in which the wood is clearly not far from the Sussex border. The Riptons are in what was clearly once well-wooded land and there can be little doubt that this *tun* takes its name from another 'wood called *Ripp.*' Jellinghaus (*Anglia*, xx. 311) followed by Moorman (*PN West Riding* 157) would associate it with *Hripum*, the early name for Ripon (Y), but seeing how early the form *Ripp* is, this does not seem very likely. (The form *Rhipp* quoted by them from BCS 161 is from a 12th cent. copy.)

The ultimate affinities of this name are obscure. It is clearly the same as the *Rip mons* which Förstemann (PN 598) associates with places called *Ripanhorst, Rippenhorst* (note the association with woodland) and with LGer *riep*, 'shore, slope,' East Frisian *ripe*, 'edge,' and a hill-name 'Auf dem *Riepen*.' These again are related to ON *rípr*, 'crag.' There can be little doubt that some word from this same stock was found in OE, for we have in the dialect of Kent and Sussex the word *ripe*, meaning 'shore, bank.' On the topographical side we may note that Ripe (Sx) is on a small but well-marked ridge rising above marshy ground and that the wood from which the Riptons seem to have taken their name must have been on relatively high ground which, a couple of miles to the north, falls to the fen-level. Another trace of the woodland which lies behind these two names is the field-name *Ripthornes*, quoted under Broughton *supra* 206 n. 1. That parish borders on Ripton. The length of the vowel in these words is not certain, and on the whole it would seem best to assume words from different grades with long and short vowels respectively.

King's Ripton was a royal manor included in the Domesday description of Hartford, Abbot's Ripton was held by the Abbot of Ramsey and later passed into the possession of the St John family. Abbot's Ripton was *Magna* Ripton (*VCH*). Corresponding to the two manors we have in 1209 (*For*) an *Abodesho* and a *Kyngesho*.

BOULTONS HUNCH WOOD (6″)

manor of Boultone 1270 AD vi

HOLLAND (Wood)

Haulund 1252 (c. 1350) Rams (p), 1301 *Ct*
Havelund 1279 RH
Havelound 1286 *For*
Haulond, Hollond 1286 *For*, 1300 *Ct*

The suffix here is ON lundr. It is defined as meaning 'grove, small wood' in EPN. Since that was written Mr Bruce Dickins has called our attention to the important rendering of *Lund* by Reginald of Durham as *nemus paci donatum*[1], which shows that this word in Scandinavianised England must have been used with the same heathen religious associations that it had in Scandinavia itself. The first element may be ON hagi, the whole name being perhaps descriptive of a sacred grove which has been 'hedged' off.

ROOKS GROVE (6″)

Roke(s)graue 1253 AD i, 1300, 1301 *Ct*
Rokysgraue 1275 *Ct*
Rokesgroue 1307 *Ct*

'Hroc's grove,' from the pers. name *Hrōc* rather than from the bird-name. This pers. name is not on record in OE but can be inferred with certainty from such place-names as Ruckinge (K) and Rockingham (Nth), and the corresponding ON *Hrókr* is well established (cf. MLR, xiv. 241). Cf. Roxton, Ruxox *supra* 64, 74.

SHOOTER'S GREEN (6″)

Sheteres dole 1297 *Ct*

'Shooter's share of the common field' (*v.* dal), *Shooter* being here presumably used as a pers. name. ME *shetere* here, as in Standard Eng, has been replaced by *shooter*, under the influence of the verb *shoot*.

[1] The passage in Reginald of Durham, c. 129 (Surtees Soc. ed. 275), is worth quoting in full: 'est ecclesia in loco qui Plumbelund dicitur...a nemoris circumcintu ita vocata, quia silvarum densissima plenitudine undique circumsepta. Qui situs loci ab Anglico proprietatis eloquio nomen propriae appellationis sortitus fuisse dinoscitur; eo quod secundum ydioma Anglicum, "lund," nemus paci donatum, cognominetur.'

WENNINGTON

Weninton c. 1000 (14th) Chron Rams 63, 1167 P
Wennitona c. 1000 (14th) Chron Rams
Wympton, Wempton 1286 *Ass*
Wenyton 1293 BM
Wenyngton 1322 FF
Wennyngton al. *Wenyngton* 1555 FF
Winnington 1766 J

Wēn- is a regular first and second element in OE pers. names (*v.* Searle) and *Wenna* seems to be an OE pet-form of these names, found in *Wennan stan* (BCS 476). Hence 'Wenna's farm' *v.* ingtun.

St Ives

ST IVES 75 H 2

S. Yvo de Slepe 1110 (c. 1350) Rams (p), 1130 BM
villa S. Yuonis 1200 FF
St Ive 1485 IpmR

The town takes its name from St Ivo whose relics were said to have been discovered here at the end of the 10th cent. For *Slepe v. infra* 222.

BROADWAY (6″)

Langebrodeweie 1217 *FF*
Bradeweye 1301 *Ct*
Brodweydych 1317 *Ct*

A street in St Ives. The name is self-explanatory. *v.* Addenda.

DARWOOD PLACE (6″)

Derhirst c. 1350 Rams

These names may possibly refer to the same place with the same development of OE deor as in Darvell (Sx).

GREEN END

Grena Sci Iuonis 1281 *Ct*

Self-explanatory.

THE HOW (6″)

le Howe 1251 (c. 1350) Rams, 1287 *Ct*

v. hoh. The name is descriptive of the site, upon a low hill-spur.

Sᴛ Aᴜᴅʀᴇʏ's Lᴀɴᴇ (6″)

> *Tawdr(e)y Lane* 1766 J, 1830 Darton

It is much to be regretted that the popular reduction of the name of the famous East Anglian saint *Æthelthryth* to *Tawdry* has disappeared in the modern genteel rendering of the name. For this reduction in the common word *tawdry*, *v.* NED *s.v.*

(Nᴇᴡ) Sʟᴇᴘᴇ (Hᴀʟʟ) (6″)

> *Slæpi* 672 (c. 1200) BCS 28
> *Slepe* 1086 DB *et passim*

v. slæp, the reference being evidently to muddy low-lying ground by the river. Slepe was the older name of the whole manor of St Ives. *v.* Addenda.

Somersham

Sᴏᴍᴇʀsʜᴀᴍ [sʌmǝsǝm] 75 F 3/4

> *Summeresham* c. 1000 Hist El
> *Sūmersham* 1086 DB
> *Sūmresham* 1086 (c. 1180) Inq El
> *Sumresham* 1086 (c. 1180) Inq El
> *Sumeresham* 1086 (12th) KCD 907, 1086 (c. 1180) Inq El, 1163, 1167 P, 1163–9 BM, 1185 P, 1228, 1229 Ch, 1230, 1236 FF
> *Someresham* 13th AD vi
> *Somersham* 1303 FA *et passim*
> *Somersam* 1549 Pat

ON *Sumarr*, OGer *Sumar* are well established as pers. names and it is probable that there was an OE *Sumor* also. This name may therefore be 'Sumor's homestead,' *v.* ham. The only difficulty is the *Summeresham, Sūmersham, Sūmresham* forms with their suggestion of a lost *m*.

If stress is to be laid on these one might take the name as from *Suðmeres-ham*, 'homestead of the south mere,' which is quite a possible name in this district. Indeed, in the bounds of the *banlieu* of Ramsey as set forth in KCD 1364 there is a *Suðmere* which may be at the very spot required. Summerfield (Nf) is from earlier *Suðmere* (cf. DB *Sutmere*), and a lost Summerfield (now Canon's Farm) in Banstead (Sr) is *Suðemeresfeld* in BCS

39, 697, 1195, and seems to mean 'open land of the south mere,' but it should be noted that the assimilation of *ðm* to *mm* is of much later date in these two cases. More probable is OE **Sunmǽr*, with later assimilation to *Summer*, hence 'Sunmær's homestead.'

CROLLODE (Fm) (6″)

aqua de Grauelode 1286 *Ass*
aqua de Grouelode 1286 *Ass*
Crowelodemare 1294 (c. 1350) *Rams*

'Crow-water-way,' probably from their frequenting it. *v.* crawe, gelad. The forms with initial *g* show the common confusion in ME between initial *c* and *g* (cf. IPN 114). It is still called Crow Lode in Dugdale, *History of Embanking* (2nd ed.), 355.

PARKHALL (6″)

Parkale 1252 (c. 1350) *Rams*
Parchalemuth 1279 RH

Here, as throughout the fen-country, one is faced with the uncertainty of the original features of the countryside. On the face of it it looks as if *Parchale* were a river-name and *Parchalemuth* the name given to some spot where it debouched into a larger stream, but we know nothing as to whether this was the case in actual fact. Apart from this we might take the name to be from OE pearroc and healh, and the whole name to mean 'nook of land marked by an enclosure.' Professor Ekwall calls attention to a *La Parrok* used of an arm of the sea in the Colne estuary in Essex and *Parrokflete* (Pat 1362), and for the interpretation of these passages notes the use of *park* in the sense 'enclosure for fish' (NED).

TURKINGTON HILL

cf. *Laurence Turkyngton in Somersham* 1551 FF

It is clear that *Turkington* here is not a place-name but a pers. name which has become attached to a property which was in possession of a man of that name. Where he came from originally we do not know, possibly it was from Torkington (Ch).

Great and Little Stukeley

STUKELEY　74 G 12/13

　　Stivecle 1086 DB
　　Stiuekele(a), Styu- 1185 P, 1318, 1320 Ch
　　Parua Stiueclai 1193 FF
　　Stifcle 1199 Cur (p)
　　Magna Steuecle c. 1200 BM
　　Stiuecl' Aristotil' 1207 P
　　Stivecle Comitis David 1220 Fees 333
　　Stiuekle, Styu-, Stiuecle 1229, 1260 FF, 1306 FF (*juxta Huntyngdone*) *et passim*
　　Stivecleya Abbatis 1252 (c. 1350) Rams
　　Steuecle 1326 Ipm
　　Stucle(y), Stukle 1362 Ch, 1386 Cl, 1433 BM
　　Steyuecle 1421 IpmR
　　Stuecle Minor and *Major* 1428 FA
　　Stewkeley 1529 FF

'Stump-clearing' *v.* styfic, leah. Cf. Stewkley (Bk) and *Styuicleage* (Mx) in O.S. Facs. iii, no. 36, 11th c. Stukeley *Comitis David* from David, Earl of Huntingdon, *Abbatis* from the holding of the Abbot of Ramsey, who had a holding in Little Stukeley. That of the earl was in Great Stukeley. Magister *Aristoteles* had a holding in Stukeley in the 13th cent. (Ramsey Cartulary i. 395) and was a clerk in the service of Henry iii.

BALDEWYNHO (lost)

　　Baldewinho, -wyn- 1279 RH, 1308 Ch, 1334 Cl
　　Baldwyneho 1286 *Ass*

'Baldwin-hoh,' from the personal name *Bealdwine*. The absence of inflexional -*es* is noteworthy.

GREEN END

　　ad Grenam 1251 (c. 1350) Rams (p)
　　Self-explanatory.

PRESTLEY WOOD (6″)

　　Prestesleye 1284 Cl
　　Prestelei boscus 1286 For
　　Prestelee, Presteley 1368 FF, 1483 AD v

OE *prēosta-lēage* (dat.), 'priests' clearing,' but why so called it is impossible now to say. This is the natural development in contrast to Priestley *supra* 74, which shows the influence of the independent word.

Upwood

UPWOOD 74 E 14

Upehude 1086 DB
Upwude 1253 BM
Upwode 1303 AD vi

'Up-wood,' in contrast to the lower woodland on the edge of the fens to the west of the village.

Wood Walton

WOOD WALTON[1] 74 E 12/13

Waltune 1086 DB, 1236 FF
Waltona 1155 BM, 1163 P (*Willelmi*)
Wauton 1225 FF, 1244 Cl
Walton juxta Sautre 1284 FF, c. 1400 Linc (*juxta Ramsey*)
Wodewalton 1300 BM
Waleton 1315 Ipm
Wallton al. *Woodwallton* 1567 FF

This is one of the Waltons in which there is a case for taking the name as from OE *weall-tūn*, 'wall-enclosure,' or 'enclosure by the wall,' rather than as OE *wēala-tūn*, 'farm of the Britons' or 'farm of the serfs,' for in the latter case one would have expected some forms with syllabic *e* (*Waleton*). One might also suggest a possible *weald-tūn* in this area (*v.* weald). There is mention of a *Waldebrigg* in this parish in 1218 (*FF*).

BARROW (lost)

Barewe 1219 (c. 1350) Rams
Barwe 1225 (c. 1350) Rams
gravam q. voc. Barwe 1279 RH

Form, and the general topography of the area indicate that here we have OE *bearwe* (dat.) from bearu, 'grove.'

[1] There is a reference to the 'wood' from which it takes its distinctive name in 1219 in Rams 168.

BEVILL'S WOOD

assartum Roberti de Beville 1219 (c. 1350) Rams

The wood and the clearing take their name from the family of *Be(y)ville* who held land here. The family probably took their name from Béville-le-Comte, dépt. Eure-et-Loir.

GAMSEY WOOD (6")

silva q. voc. Garbodeseye 1279 RH
Germeshey Wood 1566 BM

This is clearly a (ge)hæg or enclosure in a woodland district, but who its owner was is not clear. Our choice lies between taking the 13th cent. form as correct with a quite irregular development of *Garbod* to *Garmod*, or to believe that the 13th cent. form is a bad one for *Garmod*. Probably the latter is the correct solution of the name. The pers. name *Gārmōd* is not actually on record but is a perfectly regular formation. *Gārbod* is unknown, but we have *Gerbodo* in Domesday. This is a name of continental origin (cf. Forssner 106–7), and the *Gar-* might be explained as a partial anglicising of the name.

SUERSHAY (lost)

Siwardesheye 1307 (c. 1350) Rams
Suershaye 1563 FF

'Sigeweard's enclosure' *v.* (ge)hæg.

Warboys

WARBOYS 75 E 2

Wærdebusc 1077 (17th) Chron Rams
Wardebusc 1086 DB, c. 1115 *AD*, A 14399, 1163 P, 1253 BM
Wardebusche 1123–30 (c. 1350) Rams, 1167 P
Wardeboys, -bois 1148–50 BM, 1227 *Ass*, 1228 FF, 1260 *Ass*,
    1272 FF, 1321 AD i, 1348 Ch, 1350 BM, 1378 Cl
Wardboys, -bois 1189 ChR, 1428 FA
Werdebusc 1253 BM
Wardbys, Warbys 1543 BM
Wardeboisse 1556 FF
Wardboyes 1670 HMC App viii

This is a remarkable compound in which the first element is English and the second is apparently an early loan-word from the Latin *boscus*, later replaced by its French derivative *bois*. Another example of this loan-word is found in Hunts in *Wiðibusce* (*mære*) (KCD 733), i.e. withy-bush. The persistent *e* between *d* and *b* is in favour of taking the first element to be the OE pers. name *Wearda* rather than the element *weard*. For this pers. name *v.* Wardington (PN Bk 15). Hence, 'Wearda's wood.' *v.* Addenda.

CALDECOTE (lost)

Caldecote 1251 (c. 1350) Rams
Wardeboys cum Caldecote 1279 RH
Caudecote 1293 BM

v. ceald, cot.

GOLDPIT RUNNEL (6″)

Goldpitteslade 1286 For
Goldepitslade c. 1350 Rams[1]

There was evidently a small valley (*v.* slæd) here marked by a pit (*v.* pytt). Why 'gold' pit we do not know. It may be from a discovery of treasure or it may be a plant-name (cf. *Goldpyte* and *Goldenacre* as field-names in Southill (Beds) in the 13th cent.). Present-day topography affords no help.

HUMBREL'S FARM (6″)

Humberdale 1251 (c. 1350) Rams, 1255 For, 1286 *For*, 1301 *Ct*, 1346 (c. 1350) Rams
Humerdalegate 1286 *Ass*

Goldpit Runnel comes down from Humbrel's Farm and it is difficult not to think that the *slæd* and the *dale* (*v.* dæl) were parts of one and the same valley, down which a small river Humber ran, now known as Goldpit Runnel. *Humbre* is a common river-name in OE. In addition to the well-known Humber in Yorkshire we have a *Humber* (perhaps Lawern Brook, Wo) in BCS 480, a *Humbracumb* (BCS 1183) in Berkshire and two different examples of *Humbra* in Oxfordshire (KCD 714, 1296), as well as modern Humber Brooks, one a feeder of the Lugg (He) and the other in Admington (Gl), a feeder of the

---

[1] In the bounds of the *banlieu* of Ramsey (KCD 1364) this is called *Colpetslade* but the document is a late transcript and the form almost certainly corrupt.

Stour[1]. Cf. Doverdale (Wo) for a compound name of this type, and also a *torrens de Humberdale* in Milton Ernest (Bd), 1279 RH. It is clear that the name should really be *Humbrel* rather than Humbrel's Farm.

WARBOYS WOOD

    ad Boscum 1254 (c. 1350) Rams (p)

    Self-explanatory.

WOOLVEY (Fm)

    W(o)lfheye 1251 (c. 1350) Rams
    Wolneye (sic) 1291 Tax
    Wolfeye E 3 BM
    Wulveymere 1348 Cl
    Wolve c. 1350 Rams

This may be either OE *wulf-gehæg*, 'wolf-enclosure,' in which case we may compare OE *wulf-haga* (BCS 1047) or *wulf-ēg*, 'wolf-island,' with the wider sense of OE eg. In favour of the idea of 'well-watered land' is the reference to a 'mere' in 1348.

Wistow

WISTOW 75 E 1

    Kingestune id est Wicstoue 974 (c. 1350) Rams
    Wistov 1086 DB, 1114–33 BM
    Wyrstowe (sic) 1227 Ass
    Kingeston 1253 BM
    Wyn(e)stowe 1270 Ass
    W(h)ytstowe 1286 Ass
    Wistowe 1321 FF

This is OE *wīc-stow* (*v.* stow), which is variously used to denote 'house, dwelling-place or camp.' As the name here seems to be alternatively 'royal manor,' one may perhaps be permitted to take the name to mean 'site of the royal manor-house.' Wistow (Y) has the same history.

HILL (Fm)

    Monticulum $\begin{cases} de\ Wistowe \text{ n.d. KCD 1364} \\ desuper\ Wystowe \text{ 1346 (c. 1350) Rams} \end{cases}$

    Self-explanatory.

[1] Stevenson MSS.

KINGSLAND (Fm) (6″)

Kyngeslond 1252 (c. 1350) Rams

Fittingly found in a manor once known as 'king's manor.'

WISTOW WOOD (6″)

Westowode 1286 *For*

Woodhurst

WOODHURST 75 G 2

Wdeherst 1209 FF
Wdehirst 1234 FF
Wodehyrst 1252 BM

After this, forms in *hirst* (*hyrst*) and *hurst* are equally common.
The name is self-explanatory but somewhat redundant in
character, *v*. Old Hurst *supra* 211.

OCKLEY (lost)

Ocleywode 1245, 1286 *For*
Occle 1251 (c. 1350) Rams
Acle(y) 1254–67 AD i, c. 1350 Rams
Oklee 1260 *Ass*
Ocle, Okleyhil 1306 *Ct*
collem Acle c. 1350 Rams

'Oak clearing' *v*. ac, leah, appropriate enough in this parish.

WIGAN (Fm)

Wyken 13th AD i
Wyken al. *Wekyn* 1405–1538 BM
Wekyn 1535 VE
Wiggin c. 1750 Bowen

OE *wicum* (dat. pl.), 'dwelling-place, abode,' *v*. wic. Cf. Wykin
(Lei), Wicken (Nth). Cf. Eggington *supra* 122 for *k* > *g*.

Woodstone

WOODSTONE 64 J 12

Wydestun, Wudestun 973 (c. 1300–25) *Thorney* 1 *b*, 1 *a*
Wodestun 1086 DB
Wedeston 1201 Cur
Wudeston 1260 *Ass*

s

Wodeston 1268 BM
Woodston 1549 FF
Woodsone 1675 Ogilby

OE names *Wuduman* and *Ōswudu* are on record and there is evidence for other compounds such as *Wudumǣr* and *Wuduhere* (cf. Widdrington PN NbDu). From these there may have been formed a pet-form *Wud*, which might give a p.n. *Wudestun*. For the *Wyd-*, *Wed-* forms we may note the forms *wudu*, *widu* for the independent word in OE, such forms as *Widia*, *Wudga* in OE for another derivative pers. name from the same stem, and the *Wid-* forms (side by side with those in *Wud-*) discussed under Widdrington *loc. cit.*

The other and much less likely alternative is to take it as 'farm of the wood.' The normal genitive of OE *wudu* is *wuda*, but, as pointed out by Skeat, *wudes* is found in late OE.

Wyton

Wyton [witən] 75 H 1

Witune 1086 DB
Witton, Wytt- 1199 Cur, 1218 FF, 1260 *Ass*, 1287 *Ct*, 1303 *Ass*, 1307 Orig, 1535 VE
Wictun 1253 BM
Whitton 1526 LS, 1641 HMC
Witton 1766 J

This must have the same history as Witton in Droitwich (Wo), *Wictun* in BCS 361, and Market Weighton (Y), *v.* wic, tun. The meaning of the compound is obscure.

Hungry Hall Cottages (6")

cf. *Hungerton Wytton* 1270 *Ass*

Was there a 'hunger-farm' in Wyton, so called from the poverty of the land, and if so may it not possibly have survived as Hungry Hall found in this name and in Hungry Hall, just over the border in Broughton parish?

Ruddles Lane (6")

Rededale 1217 FF

If this identification is correct then the name probably means 'reed-valley,' from OE hreod, dæl, and the *s* is a mistake.

III. LEIGHTONSTONE HUNDRED

Lestune, Lestone, Delestune 1086 DB
Lectunestane 1086 DB, 1175 P
Lehtunestan 1163 P
Legtonestan 1168 P
Leochstoneston 1180 P
Wapent' de Lehtonestan 1227 *Ass*
Lettonestan 1255 *For*
Lectoneston 1285 FA
Leytoneston 1295 BM, 1303 FA
Leythtonestone 1316 FA
Leyghtonestan 1327 *SR*
Leghtonstone 1364, 1370 Cl

The *stone* at which the hundred-meeting was held is marked on Bowen's map just to the south of Leighton Bromswold, on the right side of the road, where the 1 in. Ordnance Map marks 'The Castle.' This site is central for the Hundred[1]. For a hundred-name of similar type, cf. *Bingameshou* Wapentake (Nt) which takes its name from a hoh or hill near the village of Bingham.

Alconbury

ALCONBURY [ɔ·kənbri] 74 G 12

Acumesberie 1086 DB
Alchmundesbiri, Alcmundesberia 1168, 1169 P, 1197–8 *P*
Alkmundebir, Alcmundebir, Alkmundebury 1230, 1233, 1237 Cl *et passim* to 1428 FA
Alkemund(e)bury 1238 Cl, 1252 Ch, 1287 Orig, 1299 BM, 1302 Ch, 1309 FF, 1326 Cl, 1357 Ipm
Aucmundebury 1259 Pat
Alcumbiri, Alkumbury, Alcumbury 1285 FA, 1383 Cl, 1428 FA, 1513 FF
Aumondebiry 1286 *Ass*
Alkmond(e)bury, Alcmondebury 1294 FF, E 1 BM, 1316 FA, 1356 FF
Alkemondbury 1357 Ipm

[1] This stone still exists in the village (*ex inf.* Mr C. C. Tebbutt).

Alcunbiri, Alkunbury 1303 FA, 1375 BM
Alcumdebyry 1311 FF
Alcumdeberry al. *Alcumbery* 1314 Ipm
Alkymundbery 1326 Ipm
Alkundbury 1475 BM
Aucumbury 1535 VE
Awkyngbery 1553 FF
Alcanbury 1565 FF
Awconbury, Awkenbury 1580, 1618–24 BM
Alcomberie Eliz ChancP
Aulconburye 1612 BM
Aukenbury, Aukingbury 1675 Ogilby

In the suffix we find *ber-* forms as follows: 1 in the 11th cent., 2 in the 12th, 1 in the 13th, 5 in the 14th, 2 in the 16th. The figures for *bir-* and *byr-* forms are 2 in the 12th, 8 in the 13th, 2 in the 14th, while for the *bur-* forms there are 6 in the 13th, 11 in the 14th, 4 in the 15th, 1 in the 16th and after that they prevail altogether.

'Ealhmund's burh' or 'stronghold.' Alconbury stands at the foot of a low range of hills, and the *burh* recorded in the name was probably a fortified house rather than an ancient camp. In the adjacent county of Northampton the pers. name *Ealhmund* survived into the 12th cent., and it is also found in the place-names Alcaston (Sa), Almington (St) and Alkmonton (Db).

THE GREEN (6″)

atte Grene 1327 *SR* (p)

Self-explanatory.

WEYBRIDGE (Fm)

Wardeberg 1107–13 (c. 1350) Rams
Wauberge, Wauberge nemus c. 1110 (c. 1350) Warden, 1198 *P*, 1209 *For*, 1216–30 BM, 1227 Cl, 1286 *Ass*, 1299 BM, 1301 Rams, 1343, 1378 Cl
Walberg(ia) a. 1138 RW, 1199 ChR, c. 1350 Rams
Waubergh 1227, 1247 Cl
Waberg, Waberch 1247 Cl, 1286 *Ass*
Wagheberg 1260 *Ass*
Wahberg 1282 Cl

Wauberga 1285 Cl, 1286 *Ass*, 1394 Cl
Waweberg(e) 1287, 1395 Cl
Wabrig(e) 1461, 1542 BM
Wayebrig 1470 IpmR
Waybridge 1565 BM
Wabridge 1579 FF, 1580 BM
Weybridge 1603 D

This name is the last relic of the Huntingdonshire Forest which gave importance in the early Middle Ages to the royal manor of Brampton. There can be little doubt that the name is a compound of OE weald (Anglian *wald*) and beorg, and that the whole name denotes either 'wold-hill' or 'forest-hill.' Weybridge Farm stands at the extremity of a well-marked ridge above a plain traversed by a number of streams which converge to join the Ouse near Huntingdon. This ridge is doubtless the beorg to which the name refers. The *l* was vocalised to *u* before the following (*d*)*b* and the *Waghe-*, *Wah-* forms are inverted spellings due to the normal development of *Waghe-* and *Wah-* to *Wau-*. The later developments are quite irregular.

Barham

BARHAM 74 G 10
  Bercheham 1086 (c. 1180) Inq El
  Bercham 1209 *For*, 1286 *Ass*
  Bergham 1260, 1286 *Ass*
  Berwham 1260 *Ass*
  Berewam, *Bereuham* 1279 RH
  Beruham 1286 *Ass*
  Barr(h)am 1526 LS, 1585, 1594 FF

'Hill-homestead,' aptly descriptive of the place. *v.* beorg, ham.

Brampton

BRAMPTON 74 H/J 12/13
  Brantune 1086 DB
  Branton Hy 1 (1300–25) *Thorney* 8 *b*, Hy 2 (1227) Ch, c.
    1155 *D and C Linc A* 1/1 *no.* 23, 1168 P, 12th HH, 1237
    Cl, c. 1250 MP, 1291 BM

Bramtona Hy 1 (1300–25) *Thorney* 8 *b*, c. 1150 *D and C Linc
A* 1/1 *no.* 8, 1168 P, 1242 Cl
Brampton 1227 *Ass (et passim)*
Braunton 1241 Cl
Brampton juxta Huntyngdon 1343 FF

'Brier' or 'bramble farm' *v.* **brame, tun.**

BRAMPTON WOOD

boscus de Brampton 1219 *For*
boscus de Brompton 1286 *For*

BROMHOLME BRIDGE (6″)

Bramholm 1253 AD ii
Bromholm 1327 Ch

The bridge is on a small tributary of the Ouse by land marked
'liable to flood,' so that it is clear that here we have ON holmr
in the sense 'low-lying ground by a river or stream' which it
develops in English dialect. The first element shows the common confusion of **brame** and **brom**.

HARTHAY [ha·ti]

Hertehey 1219 *For*
Hertheie 1227 ClR
Herteye 1227–9 Ch
Herthey 1279 RH, 1298 Ipm, 1307 (c. 1350) Rams
Harthey 1299 BM
Harty 1836 O

'Stag enclosure' *v.* **heorot, (ge)hæg.** It may safely be
assumed that the place originally formed part of Weybridge
Forest. Harthay is less than two miles from Weybridge Farm.

PORT HOLME

Portholme 1417 FF

As this borders on the town of Huntingdon it can safely be
interpreted as 'low-lying ground' (*v.* **holmr**) by or belonging to
the port or town, cf. Portmeadow in Oxford. Cf. *Huntendunport*
in the forged Peterborough Charter, contained in the ASC (E)
s.a. 656.

WOODHOUGHTON (lost)

> *Wodehouton* 1219 *For*
> *Brampton cum Wodehoghton* 1286 *Ass*

In each case the place is associated with Brampton, and as Houghton is the other side of Huntingdon the reference can hardly be to that place. Probably this 'Wood Houghton' (*v.* hoh, tun) was so called to distinguish it from the other Houghton down by the Ouse.

Brington

BRINGTON 74 G 8/9

> *Breninctune* 1086 DB
> *Brincthon* 1252 BM
> *Bryninton, Brininton* 1253 BM, 1259 AD i, 1260 FF, 1285 FA, 1344 BM
> *Brinington, Brynington* 1271 AD i, 1286 *Ass*
> *Brimington* 1279 RH
> *Brimpton, Brympton* 1291 Tax, 1428 FA
> *Brunnington* n.d. *AD* A 3562
> *Brington* 13th AD i, 1267 Cl, 1443 BM
> *Bringtun* c. 1350 Rams
> *Brynkton* 1360 Cl
> *Bryncton* al. *Brynton* 1594 FF
> *Brington* or *Brynton* c. 1750 Bowen

This is clearly OE *Bryningtun*, 'Bryni's farm,' *v.* ingtun. The ancient pers. name *Bryni* occurs also in the place-names Brimpton (Berks), Brington (Nth), identical with the present name, Briningham (Nf), and Bringhurst (Lei), and probably in Burniston and Brinsford (Y).

Buckworth

BUCKWORTH 74 F/G 10/11

> *Buchesworde* 1086 DB
> *Buckeswirtha* c. 1150 (13th) Hugonis Candidi Historia
> *Buckeswrda* 1180 BM
> *Buckesworth* 1220 Fees 334, 1294 FF (p)
> *Bukesw(u)rth* 1223 Bract, 1247 Pat, 1248 FF, 1282 Ipm, 1285 FA

Buckew(o)rth 1225 FF, E i BM
Bokewurth 1227 *Ass*
Bokesw(o)rth 1242 Fees 922, 1267–85 AD i
Bucwurth 1243 Cl
Bockysworth 1276 RH
Buckeswrþe 1279 RH
Bockeworth 1294 FF
Bokworth 1303 FA, 1428 FA
Buckworthe 1316 FA, Eliz ChancP
Bukworth, Bucworth 1327 *SR*, 1341 Cl, 1351 Ipm, 1353 Cl,
    Ipm, 1369 Cl, 1370 FF, 1428 FA

'Bucc's enclosure' *v.* worþ. For this pers. name, *v.* MLR
xiv. 236.

Bythorn

BYTHORN 74 G 8

Bierne 1086 DB, c. 1350 Rams
Bi-, Bytherne 1127 (c. 1350) Rams, 1248 FF, 1253 BM, 1259
    AD i, 1260 *Ass*, 1357 Ipm (*by Elyngton*), 1437 FF, 1451 BM
Byern(e) 1252 BM, 1286 *Ass*
Bithorne, Bythorn 1285 FA, 1545 FF

'By the thorn-bush' *v.* bi, þyrne.

CLACK BARN (6")

terra quondam Clack c. 1350 Rams

This land is named from a man bearing the common Scand.
pers. name *Klakkr*, which was early adopted into Anglo-
Scandinavian personal nomenclature, and appears in the 12th
cent. in the forms *Clac, Clach*.

Catworth

CATWORTH [kætəθ] 74 H 9

Catteswyrð (dat.) 972–92 (c. 1200) BCS 1130
Parua Cateuuorde, alia Cateuuorde 1086 DB
Cadeuurde, Cadeworde 1086 (c. 1180) Inq El
Catteswurda 1163 P
Cattewurda 1167 P
Catew(o)rth 1199 Cur, 1248 FF, 1287 *Ass*

Magna Cattewrth c. 1200 BM
Cattewurth 1224 FF, 1227 *Ass*, 1231, 1272 FF
Catteworth 1277 Cl, 1285 FA, 1287 *Ass*, 1290 Cl, 1302, 1428
   FA
Katesworth 1286 QW
Catw(o)rth 1288 Cl, 1325 Fine

'Catt's enclosure.' The existence of a pers. name *Catt* in OE
is made certain by such names as Catshall (Sf), Catsfield (Sx),
Catshill (St, Wo), and further evidence will be found in MLR
xiv. 237. The *s* here and in Buckworth *supra* 235 would readily
be lost owing to the existence of weak forms *Catta*, *Bucca*. Cf.
also Catwick (Y).

Coppingford

COPPINGFORD 74 E/F 11
   Copemaneforde 1086 DB
   Coupmanneford 1146 *Cott* vii. 3
   Copmanesford 1207 Abbr
   Copmaneford 1225 Pat, 1248, 1253 FF, 1285 FA
   Copmanneford 1227 *Ass*, 1273 FF
   Copmanford E 1 BM, 1303 FA, 1308 FF, 1327 *SR*, 1351 Ipm,
      1353 Cl, 1362 FF, 1372 BM, 1383 Cl, 1428 FA, 1444 IpmR,
      1501 Ipm
   Coupmanford 1286 *Ass*
   Copmansford 1290 Misc
   Copemanford 1316 FA
   Copmandesford 1382, 1389 FF
   Coppemanford 1428 FA
   Coppyngford 1535 VE, 1564 FF
   Copmanford al. *Coppingford* 1584 FF

This is a curious hybrid name. The first part is clearly Late
OE *Coupmanna* from ON *kaupmanna*, 'of the traders,' and the
whole name means 'traders' ford.' Cf. a similar *chypmanna ford*
in Wiltshire (BCS 879) from OE *cīepmann*, the corresponding
native word. The ford must be that to the west of the village,
rather less than a mile away. This ford does not seem likely ever
to have been a very important one. A cart-track leads down to
it and it is continued on the other side as a footpath which leads

on by roads and cart-tracks through the Giddings to the valley of the Nen at Warmington. It is tempting to think that these 'merchants' used not the line of track just indicated but the ancient and well-known Bullock Road which branches off from Ermine Street, just to the south-west of Coppingford, skirts the village itself, and makes its way in a fashion closely similar to that of the road just described, to the Elton-Chesterton road, whence either the Nen could be crossed at Elton or the Ermine Street rejoined at Alwalton. The two roads run almost parallel throughout their course and it is just possible that both alike represent ancient tracks of importance and that we should be wrong in identifying the merchants with the Bullock Road alone. The fact that Henry of Huntingdon, the historian, held, as archdeacon, a local chapter at Coppingford (Croyland Cartulary) shows that the place was easy of access in the first half of the 12th cent.

COPPINGFORD WOOD

    boscus de Copman(e)ford 1248, 1286 *For*

Covington

COVINGTON 74 H 8

    Covintune 1086 DB
    Couyngton 1260 *Ass*, 1331 FF, 1493 Ipm
    Couinton 1272 FF, 1279 RH, 1285 FA, 1303 *SR*, FA
    Couenton 1478 FF

'Cofa's farm' *v*. ingtun. Cf. Covenham (W) and Coveney (C). We have the strong form of this pers. name in Cosgrave (Nth), DB *Covesgraue*.

Easton

EASTON [iˑsən] 74 H 10

    Estone 1086 DB *et passim*
    Eston al. *Esson* 1578 FF

'East farm,' presumably in contrast to Old Weston on the other side of the hundred-centre at Leighton. Note the contrast of *Esson* here and *Wesson* in the forms for Old Weston *infra* 250.

CALPHER WOOD

> *boscus de Calfho* 1271 *FF*
> *Calfo* 1279 RH

'Calf-hill' *v.* cealf, hoh. The wood is on the side of a hill but a concave one rather than a convex one such as is usually called a hoh. There is a hoh in the usual sense just to the north-west of the wood and one cannot of course be sure from exactly which spot the wood took its name.

Ellington

ELLINGTON 74 H 11

> *Elintune* 1086 DB, 1106–13 BM
> *Elint(h)on, Elynton* 1207 *P*, 1227 *Ass*, 1228, 1241 FF, 1253 BM, 1285 FA
> *Elindon* 1228 FF
> *Elyngton, -ing-* 1267–85 AD i, 1286 *Ass*, 1327 *SR*, 1346 Cl *et passim*

'Eli's farm' *v.* ingtun. For this name, *v.* Redin 134. The complete absence of forms with double *l* forbids our assuming that we have the more usual pers. name *Ella*[1].

CALDECOTE (lost)

> *Caldecote* 1279 RH, 1330 Pat (*under Brouneswold*)

v. ceald, cot. Distinguished as 'under Bromswold.'

COTON BARN (6″)

> *Cotene* 1286 *Ass*, QW
> *Coten* 1292 BM

'(At the) cottages,' from the dat. pl. of cot.

ELLINGTON THORPE

> *Sybethorp, Sib-* 1227 ClR, 1279 RH, 1298 Ipm, 1322 AD i
> *Sibbetorp* 1227 *Ass*, 1255 *For*
> *Sybetorp* 1236, 1241 FF
> *Sibbethorp* 1258 FF
> *Elyngton cum Sibethorp* 1286 *Ass*

---

[1] The forms which Skeat gives for this name with medial *dl* and *tl* belong to Elton and not to this place.

Sibthorpe 1316 FA, 1357 Ipm
Sibythorpe 1323 AD i
Sipthorp 1571 FF
Thorp or *Sibthorp* c. 1750 Bowen

'Farm of Sibba (OE) or Sibbi (ON).' It is difficult to say here whether this name contains the English or Danish þorp. The complete absence of any *throp* forms is somewhat in favour of the latter alternative, and this is confirmed by the existence of another *Sibbethorp* in the northern Danelaw (now Sibthorpe (Nt)). The later dropping of the name of the owner and the assimilation of the name to a type common in the Danelaw, in which *Thorpe* is added to the name of a parent village, is to be noted. Cf. Mattersey and the adjacent Mattersey Thorpe (Nt).

Great and Little Gidding

GIDDING 74 E 10

Redinges 1086 DB
Gedelinge 1086 DB
Geddinge 1086 DB, 1147 BM, 1166 P, 1193, 1208 FF, 1227
    Ass, 1237 FF, 1252 FF (*Magna*), 1276 RH, 1316 FA, 1340
    Ipm, 1385, 1399 IpmR
Geddinges 1168 P (p), 1207 P
Gedinges 1185 (c. 1200) *Templars*
Guedding 1198 Fees 9
Gydding 1253 FF, E 1 BM (*Parva*), 1297 Ipm, 1304 *SR*
Geddingg Prioris 1276 RH, (*Engayne*) ib.
Gidding, -yng 1285 FA, 1304 Ch, 1316 FA, 1323 Ipm, *SR*,
    1328 Ch, 1341 FF *et passim*
Great Gedyng 1549 Pat

'Gydel's people' *v.* ingas. Of the three DB forms, *Redinges* is clearly a scribal error resulting from the confusion of the late OE letters *g* and *r*. The *Gedelinge* form cannot be dismissed so briefly. It occurs in the record of disputed claims to land which forms an appendix to the Hunts DB and it so happens that, occurring at the end of a line, it is divided *Gede-linge*. It is difficult to see how this can be an example of scribal confusion between *dd* and *dl*, quite apart from the fact that this confusion, very common in later records, is not likely to occur in 11th cent.

handwriting. It is therefore probable that Gidding contains a pers. name of diminutive type, *Gyd(e)la* rather than the *Gydda* to which Ekwall refers in his discussion of this name (PN in *-ing* 74, 88, 160). Great Gidding is Gidding *Prioris*, so called from the holding of the Prior of Huntingdon (FA ii. 472), Little Gidding is Gidding *Engayne* (FA ii. 470, 471, 475) (*VCH*).

Steeple Gidding

STEEPLE GIDDING

Stepelgedding 1260 *Ass* (p)
Geddingg Abbatis 1276 RH
Stepel Guiddyng 1291 Tax
Gydding Abbatis 1294 BM
Steple Gyddyng al. *Abbott Gyddyng* 1598 FF

'Abbot's' from the holding of the Abbot of Ramsey (*VCH*). At the present time there is a tower and spire at both Great and Steeple Gidding. One must presume that at one stage in their history Steeple Gidding alone had this distinction.

Grafham

GRAFHAM 74 J 11

Grafham 1086 DB, 1140–5 BM, 1159 P (p), John BM, 1200 FF, 1235 Cl, 1237 FF, 1278 *Ass*, 1302 Cl, 1303 FA, 1314, 1323, 1348 Ipm, 1373, 1400, 1416 FF, 1428 FA
Grafam 1140–50 BM
Graffham 1207 FF, 1236 Cl, 1316 FA
Grapham E 1 BM, 1285 FA
Grofham 1342, 1368 Cl, 1370 FF, 1382 Cl, 1409 BM, 1411 FF, 1428 FA, 1434, 1460 FF, 1548 Pat
Groff(e)ham 1387 IpmR, 1509 FF, 1535 VE, 1549 Pat, 1575 FF
Grofom 1393 BM
Gropham 1526 LS
Graff(e)ham 1535 VE, 1554 FF
Groffam 1535, 1553 FF
Croffham al. *Groffham* 1600 FF

'Grove-homestead' *v.* graf, ham. Grafham, three miles from Weybridge Farm *supra* 232, is in an old forest-area, and in the

Ramsey Cartulary (ii. 304) we have mention of *sex gravetae* in Grafham. In the history of the name there has been a curious fluctuation between forms with short *o* from the long *o* which developed normally from OE *ā*, and forms with short *a* which developed before the rounded *o*-forms arose.

EAST PERRY

> *Peri* 1147 BM
> *Pirie.* c. 1180 (c. 1230) *Warden* 21 *b*
> *Est Perye* 1323 Ipm
> *Pirie in Grofom* 1393 BM

'Pear-tree' *v.* pirige.

Hamerton

HAMERTON 74 F 10

> *Hambertun* 1086 DB
> *Hamertun* 1152 BM, c. 1155 (13ᵗh) Colchester
> *Hamereton* 1168 P (p)
> *Hamerton* 1199 Cur, 1219 FF, 1227 *Ass* (*et passim*)
> *Hammerton* 13th AD i, 1587 FF

It is difficult to believe that if there really was a *b* in this name in the time of DB that it would have then completely disappeared. Rather we must take it to be a sporadic appearance of a common type of epenthetic *b* between *m* and *r* (cf. Campton *supra* 167). That leaves us with *Hamer-* as the more correct form of the first element. Skeat suggests that there may have been an OE pers. name *Hamor*, as illustrated by such a p.n. as Hameringham (L), and this is endorsed, somewhat hesitatingly, by Ekwall (PN in -*ing* 141). Cf. OGer *Hamar*, ON *Hamarr*. If we believe we have this name here we must take the name to be one of that rare type in which the pers. name and the second element are placed together without any connecting genitival inflexion (cf. Kimbolton *infra* 243). The only other alternative is to take the full first element as having been a plant-name such as *hamor-secg*, 'hammer-sedge,' or *hamor-wyrt*, 'hammer-wort' or 'black hellebore' (cf. Skeat PN Hu 344) and to explain the present form as due to that dropping of the middle element from a triple p.n. compound for which Ritter gives a good deal of evidence (88 ff.). Hamerton stands on low ground, by Alconbury Brook.

DIPSLADE COPPICE (6")

    cf. *Depeslade* (*in Stilton*) 1241 *FF*

This must have the same history as the lost place in Stilton and mean 'deep valley,' *v.* deop, slæd. There is a well-marked depression here.

Keyston

KEYSTON *olim* [kestən] 74 G 7

    Chetelestan 1086 DB, 1163, 1166 P
    Ketelestan 1172 P, 1209 *For*
    Ketillistan 1173 P
    Ketstan 1227 *Ass* (p)
    Ketlestan 1227 *Ass*
    Ketelston 1248 *For*
    Keston 1255 *et passim* to 1442 IpmR
    Ketston 1260 *Ass*
    Kestan 1260 *Ass*, 1272 FF, 1286 Orig, 1293 Ipm, 1299 FF,
        1303 *SR*, FA, 1428 FA
    Kesston 1309 Ipm
    Kaiston 1526 LS
    Keyston 1553 FF
    Kayston al. *Keyston* 1560 FF
    Keston 1647 HMC App viii. 2
    Keyston or *Keston* c. 1750 Bowen

'Ketill's stan.' The pers. name is of Scand. origin. As Keyston is the most westerly village in the county it is probable that the *stan* was a boundary stone marking the westernmost limit of the shire or of some more ancient local division which it represents.

CROW'S NEST HILL

    cf. *Krowenisthill in Ellington* 1216–30 BM

A clear duplication of name in parishes not so very distant.

Kimbolton

KIMBOLTON *olim* [kiməltən] 74 J 9

    Chenebaltone 1086 DB
    Kenebolton 1185 Rot Dom

Kinebald' 1232 Cl
Kinebalton, Kyn- 1232 Cl, 1260 *Ass*, 1288 Cl, 1293 Ipm
Kenebauton 1232 Cl, 1236 Pat, 1303, 1316 FA, 1368 AD i
Kenebavdton 1247 *Ass*
Kenebalton 1260 *Ass*
Kynebauton, Kine- 1284 FA, 1289 *Ct*, 1292 NI, 1293 Ipm,
    1302 Ch, 1303 FA, 1310 FF, 1377 Cl, 1428 FA
Kynebauton cum Soke 1327 *SR*
Kymbalton 1329, 1361 Fine, 1362, 1373 Cl, 1427 FF, 1443
    HMC App x. 4, 1542 BM, 1585 AD v
Kimbolton, Kym- 1330 Ch, 1428 FA
Kymmolton al. *Kymbalton* 1544, 1553, 1575 FF
Kymbolton al. *Kynebauton* 1572 FF
Kymolton 1588 AD v

'Cynebeald-farm,' with immediate juxtaposition of the pers.
name and the second element. *v.* tun and cf. Wymington
supra 50.

BRIGHTAMWICK (lost)
  Brihtelmewic 1167 P
  Brihtelinwic (sic) 1279 RH
  Brythomwyk, Brythamwyk 1347 AD ii
  'Byrhthelm dairy-farm' *v.* wic.

DUDNEY WOOD (6″)
  boscus de Dudenhey 1241 *FF*
  'Duda's enclosure' *v.* (ge)hæg.

KIMBOLTON PARK
  Parcus de Kimbolton 1248 *For*

NEWTOWN
  Neuton 1286 *Ass*
  Newtowne Eliz ChancP
  'New farm' *v.* niwe, tun.

STONELY *olim* [stɔnli]
  Stanlegh 1260 *Ass*
  Stonle(gh) 1260 *Ass*, E i BM, 1279 RH *et passim*

Staynley 1287 *Ass*
Stonley 1766 J

'Stony clearing' *v.* stan, leah. In the modern form the vowel has been lengthened under the influence of the independent word as in the identical Stoneleigh (Wa).

WORNDITCH [wəˑnditʃ]

Wormedik 1279 RH
Wormedich, -ych 1279 RH, 1286 *Ass*, 1360 AD iii, 1366 AD i, 1432 AD i, 1535 VE, 1579 FF
Wermedych 1286 *Ass*, 1288 Ipm
Wyrmedych cum Neuton 1286 *Ass*
Wyrnedich 1286 *Ass*
Warmedyche 1587 FF

'The ditch or dike of *Wyrma* or possibly of *Wurma.*' *v.* dic. The old forms may perhaps be explained from *Wyrma* alone for OE *y*, ME *u*, is occasionally represented by *o*. But as OE *y* in Hunts usually appears as *e* or *i* we should perhaps take the *Worm-* forms as from OE *Wurma*. The name *Wyrma* is found in Warmington (Nth), Wormegay (Nf) and Worminghall (Bk).

Leighton Bromswold

LEIGHTON 74 G 9/10

Lectone 1086 DB, 1229, 1237 Cl
Lehtone 1227 *Ass*
Letton 1230 FF
Leghton 1253 FF
Lethton 1260 *Ass*

BROMSWOLD

Bruneswald 1168 P (p)

The two names are first associated in

Leghton super Bromeswolde 1287 *Ass*

and similar forms with the second element as follows

Brouneswald 1286 QW
Brouneswold 1287 *Ass*, 1347 FF
Broneswold 1300 *Winton*

T

Brunneswold 1301 FF

and then

Leyghton Brounsold 1542 FF
Leyghton Bromeswold al. *Brymeswold* 1549 FF
Leyghton Bromsolde 1554 FF

v. leac, tun. The name must have been used of a farm well stocked with garden-produce. The second name is from OE *Brunes-w(e)ald*, 'Brun's wold,' and describes the high open ground on which Leighton stands. *v.* weald.

SALOME WOOD etc. [sɔləm]

Salne 1227 *Ass* (p), 1260, 1286 *Ass*, 1364 Ch
Sale 1227, 1260 *Ass*
Graua de Salem 1248 *For*
Salene 1279 RH, c. 1350 Rams
Salue (sic) 1284 Cl, 1286 QW
boscus de Salnee 1286 *For*
Salewe 1316 FA
Leyghton cum Salene 1327 *SR*
Salengrove 1357 Ipm
Leighton Salon 1466 IpmR
Salom Wood 1610 Speed

OE (*æt þæm*) *sēalum*, 'at the willows,' *v.* sealh. The modern final *m* and *e* are quite misleading.

Molesworth

MOLESWORTH *olim* [mʌlzwəθ] 74 G 8

Molesworde 1086 DB
Mulesw(u)rth 1208 FF, 1227 *Ass*, 1228, 1231 FF, 1279 RH, 1285 FA
Moleswurth 1232 Cl
Mulesworth 1234 Cl, 1316 FA
Mulisw(o)rth 1248 FF, 1289 Cl
Mollesw(o)rth 1253 FF, 1286 *Ass, For*, 1327 Ch, 1357 Cl, 1359 Ipm
Mullesworth 1260 *Ass*, 1303 *SR*, 1325 FF, 1327 *SR*, 1330 Ipm, 1334 FF, 1386 Cl, 1389 IpmR, 1393 Cl, 1428 FA, 1429, 1467 FF
Molesworth 1303, 1428 FA

Moulisworthe 1324 BM
Mellesworth 1353 Cl, 1358 Ipm
Mullysworth 1407 FF
Mowlesworth 1467 BM, 1571 FF
Mowlesworth al. *Mullesworth* 1515 FF
Mulsworth 1535 VE
Moulsworth 1675 Ogilby

'Mul's enclosure' *v.* worþ. It is evident that there was a long struggle between forms in which OE *Mūl* underwent its natural shortening to *Mull* in the compound and those in which the long vowel was kept. The latter ultimately prevailed here, as also in Moulsford (Berks), Moulscombe (Sx), Molesey (Sr), Moulsham (Ess). A similar struggle took place with regard to Moulsoe (Bk).

Spaldwick

SPALDWICK [spɔ·ldik] 74 H 10

Spalduice 1086 DB
Spalduuic(c) 1086 (c. 1180) Inq El
Spaldewic(k) 1086 (c. 1180) Inq El, 1155 (1329) Ch, 1163, 1167, 1185 P
Spalduic 1109 BM
Spaudewyk 1209 *For* (p)
Spaldewyk,-wik 1260 *Ass*, 1285 FA, 1327 *SR*, 1330 FF, 1428 FA
Spaldinwike 1286 *Ass*
Spaldingwik 1286 QW
Spaldwycke 1316 FA
Spaldewick 1326 Cl
Spaldwyk 1554 BM
Spaldicke 1583 FF
Spalwick 1610 Speed

It is impossible to carry the story of this name any further than Ekwall (PN in *-ing* 88–9), where he relates it to Spalding (L), Spaldington (Y), Spalford (Nt), earlier *Spaldesford*, Spalding Moor (Y) and the *Spalda* of the Tribal Hidage (BCS 297). The natural thing would perhaps be to relate all these names to some lost pers. name, but there is no other evidence for such in the Germanic languages, and Ekwall suggests that we have a lost river-name in Spalford, Spaldwick and Spalding, while in Spalding Moor and Spaldington we may have an actual settlement from Spalding itself. This river-name he takes to be derived either from OE *spald*, 'spittle, foam,' or an OE *spald, cognate with OGer *spalt*, denoting 'trench, ditch,' and specially applicable to the fen-land rivers. It should be noted in addition that the name in the Tribal Hidage refers almost certainly to the Spalding district, and as at least two other names in the Hidage are taken from river-names (cf. *Gifla* = Ivel and *Hicca* = Hiz) this identification slightly strengthens the case for a river-name *Spald*.

UPTHORPE

> *Upthorp(e)*　1260 *Ass* (p), 1269 FF, 1279 RH
> *Opþorp*　1279 RH
> *Stowe cum Upthorp*　1286 *Ass*

'Upper farm,' because it is on higher ground away from the valley in which Spaldwick itself lies. For *thorpe* here, cf. Ellington Thorpe *supra* 239.

Stow

LONG STOW　74 H 9

> *Estou, Estove*　1086 (c. 1180) Inq El
> *Stowe*　1219 FF
> *Oueristowe*　1248 FF
> *Longestowe*　1286 *For, Ass*, 1327 Cl
> *Ourestowe*　1286 *Ass*
> *Langestowe*　1344 Fine

v. stow. We have no knowledge of the sense in which the term was used in this particular case. *Long* probably because the village is somewhat long and straggling, *Over* because it is on some of the highest ground in the neighbourhood.

STOW GROVE (6″)

> *Stowegrove*　1279 RH

Tilbrook[1]

TILBROOK　74 J 8

> *Tilebroc*　1086 DB, 1202 Ass
> *Tillebroc*　1202 Ass, 1203 FF, 1227 Ass, 1242 Fees 888
> *Tillebrok, Tyll-*　1227 Ass, 1276 *Ass* (*et passim*)
> *Tylebrok, Til-*　E 1 SR (*et passim*)

---

[1] Formerly in Beds.

Tylbroke 1276 *Ass*
Tilebroke 1287, 1307 *Ass*

'Til(l)a's stream' *v.* broc. If that is the case this must have been the old name of what is now called the river Til or (further down stream) the river Kym. Alternatively we might perhaps take broc here to mean low-lying land (*v.* EPN). The river-name *Til* is certainly only a back-formation from Tilbrook.

HARDWICKS

Herdwik E 1 *SR*

v. heordewic.

Upton

UPTON 74 F 11

Opetune 1086 DB
Upton 1285 FA, (*juxta Wodeweston*) 1286 *Ass*

'Upper farm,' either from its general situation or (with Weston) from their relation to Alconbury in particular.

STANGATE HILL

Stangate 1146 Cott vii. 3
regalis via q. voc. Stangate 1286 *Ass*

Gate here must be the Scand. loan-word *gate* from ON gata, as the name seems to have been actually applied to Ermine Street itself. It is borne by this road at the point where it climbs to cross the ridge between the fen-land east of Sawtry and the flats along Alconbury Brook. Stukeley (*Itinerarium Curiosum*, Iter v, 81) says that the name means 'paved with stone,' some of it still being in existence near Stilton in his day, i.e. 1742[1].

UPTON WOOD

boscus de Upton 1227 *Ass*

Alconbury Weston

ALCONBURY WESTON 74 F 11

Westune[2] 1086 DB
Alkmundebir Weston 1227 *Ass*

---

[1] This reference is due to the kindness of Mr E. C. Gardner.
[2] Has *soke* in Alconbury.

Wodeweston 1260, 1286 *Ass*
Awconbury cum Weston 1557 FF
Aukingbury Wiston 1675 Ogilby

'West-farm,' apparently in relation to the settlement at Alconbury, *Wood*-weston in contrast to Old or *Wold* Weston *infra*, cf. also Old Hurst and Woodhurst *supra* 211, 229.

HERMITAGE WOOD

'*a place called Hermitage*' 1326 Ipm

Old Weston

OLD WESTON

Westune 1086 DB
Wald Weston 1227 *Ass*
Wold(e)weston 1249 Misc, 1260 *Ass*, FF, 1305 FF, 1316 FA, 1343 AD v, 1346 Cl
Weston juxta Leytthon 1274 FF
Weston de Wald 1276 RH, 1286 *For*
Weston de Waldis 1285 FA
Weston super Waldis 1303 *Ass*, FA
Weston super Brouneswold 1442 FF
Oldweston 1535 VE
Olde Weston 1553 FF
Weston al. *Owld Wessen* 1594 FF

'West farm' in relation perhaps to Easton (*v.* 238 *supra*). It stood on the *wold* known as Bromswold (cf. Leighton Bromswold *supra* 245) and came to be known as Wold Weston in contrast to Alconbury or Wood Weston.

COCKBROOK LODGE

aqua de Cukusbrygg, Cukusbrygg, Cukisbrygg 1286 *Ass*

The association of this stream in the Eyre Roll with Winwick and Old Weston makes it practically certain that here we have a name connected with the Cockbrook found in this name and in Cockbrook Farm, Lane and Spinney in the neighbouring Brington parish. The brook is the now nameless stream which flows by Old Weston and the bridge must be the one which crosses it at that village. The etymology is obscure.

OLD WESTON GROVE (6″)

Westongraue super Waude 1261 FF

Winwick

WINWICK 74 E 9

Wineuuiche 1086 DB

Winewic, Wynewik 1195 BM, 1227 *Ass*, FF *et passim* to
1428 FA

Wynwyk 1348 Ipm, 1428 FA

Wynnewyke 1393, 1447 IpmR

Wynnewyg 1399 IpmR

Windwike 1535 VE

Winnick 1641 HMC

'Wina's dairy-farm' *v.* wic.

Woolley

WOOLLEY 74 G 11

Ciluelai (sic) 1086 DB

Wulueleia 1158 P

Wulfelea 1180 P

Wluele 1219 *For*

Wolvele 1220 Fees 334, 1285 FA, 1315 Ipm

Wulvele 1227 ClR

Wlfle(g) 1242 Fees 922, 1248 FF

Wolfle 1255 *For*

Wullegh 1260 *Ass*

Wolle(y) 1276 RH, 1303 FA, 1315 Fine, 1580, 1627 BM

Well(e)y 1314 Orig, 1316 Fine, 1317 Cl

Wulle(e) 1390 FF, 1399 IpmR

OE *wulfa-leage* (dat.), 'wolves' clearing,' a fairly common
p.n. *v.* leah.

IV. TOSELAND HUNDRED

Toleslund 1086 DB

Touleslund 1227 *Ass*, 1303 FA

Touleslond 1316 FA, 1364, 1370 Cl

Tousland 1428 FA

Toseland 1585 D

The name is of pure Scandinavian origin (*v.* Toseland *infra* 272), and it is a matter of interesting speculation whether one should believe in any definite connexion between the religious associations of the old word lundr (cf. Holland *supra* 220) and the ceremonial of the meeting of the Hundred. A large stone, which might have been the base of a cross, is in the churchyard, and is locally said to have been the 'moot-stone.' Part of the fragmentary Roman road from Sandy to Godmanchester is known in the neighbourhood of Toseland as 'Moats' or 'Moots Way'[1].

Abbotsley

ABBOTSLEY [ævəzli] *olim* [ɔˈbzli] 84 D 13

Adboldesl' Hy 2 (Hy 3) *St Neot* 81
Albedesleg 1257 FF
Adboldeslee Hy 3 *St Neot* 31
Abbodesle 1270 FF, 1303 FA, 1313 Fine, Cl
Albodeley 1272 FF (*Magna*)
Alboldesle(g) 1279 RH, 1310, 1334 FF
Albodesley, Addeboldeslee 1286 *For*
Albotesle(y) 1286 QW, 1300, 1315 FF, 1318 Cl, 1436, 1470 FF
Abbotesle(y) 1286 QW, 1318 Ipm, Cl, 1381 Cl, 1428 FA
Abbottesle(y) 1316 FA, 1517 FF
Albodesle 1317 FF
Abboltesle(y) 1375 Cl
Aubottesley 1549 Pat
Albottysley al. *Aubsley* al. *Abbottysley* 1561 FF

'Ealdbeald's clearing' *v.* leah. The later forms show the influence of the common word *abbot*. Abberley (Wo) has the same history and in 1535 (VE) has alternative forms *Abotesley* and *Aburley*.

Buckden

BUCKDEN [bʌgdən] 74 J 12

Bugedene 1086 DB, 1185 P, 1238 Cl, 1286 *Ass*
Bugendena 1155–8 (1329) Ch, 1185 P

[1] *ex inf.* Mr R. C. Gardner.

Buggeden 1167 P, 1227 Ch, *Ass*, 1235, 1238 Cl
Buggenden 1167 P
Bokeden 1245, 1255 *For*, 1324 Cl, 1327 *SR*, 1337 Cl, 1340 FF, 1354 Ipm, 1365 BM
Bogeden 1245 *For*, 1286 Dunst
Bukeden 1248 FF, 1260 FF, 1270 *Ass*, 1272 FF, 1286 *Ass*, 1316 FA, 1329 Ch, 1365 BM, 1497 FF
Bukedon 1270, 1286 *Ass*
Buckden 1279 Cl
Bokedon 1286 *Ass*
Buckeden 1335, 1376 Cl
Bukkeden 1401 BM
Bukton 1409 IpmR
Bugden 1478 FF, 1535 VE, 1558 FF, 1574, 1615–35 BM, 1660 *Linc*
Bukden, Bucden 1485 FA, 1526 LS
Bugden or *Buckden* c. 1750 Bowen

'Bucge's valley' *v.* denu. The later unvoicing of *Bug-* to *Buck-* is perhaps due to the influence of the not very distant Buckworth. *Bucge* is one of the most ancient hypocoristic feminine names found in OE. It may represent either one of the numerous feminine compound names ending in *burg* (e.g. *Eadburg*) or a feminine name such as *Burghild*, in which this name forms the first element. Cf. *Buggildstret, Bucganstræt* (BCS 126, 1201) as alternative names for Buckle Street (Wo). The name *Bucge* occurs also in Bognor (Sx), *Bucgan ora* BCS 50. No masculine *Bugga* is known.

EDWOLDESHEY (lost)

Edwaldeshell' (sic) 1200 Cur
Edwaldessay 1227 *Ass*
Edwoldeshey 1279 RH (*boscus*)
Eldwoldeshey 1307 Rams

'Eadweald's enclosure' *v.* (ge)hæg.

HARDWICK

Bukeden cum Herdwyk 1286 *Ass*

v. heordewic.

STIRTLOE

> *Stirt* 1209 *For*
> *Stert* 1231, 1284 FF
> *Sterth* 1257 BM
> *Stertelawe* 1286 *Ass*
> *Stirtloe* 1574 BM

This must be a compound of **steort** used of a 'tongue or tail of land' and hlaw, 'hill.' Cf. *Sturteslowe* in Ravensden (1269). It is curious that there are no signs of what would have been the normal development to *Start* and *Startlow*. Mr R. C. Gardner informs us that there is nothing but a slight ridge here, rising up from the Ouse, nothing that we should really call a hill.

Diddington

DIDDINGTON 84 A 12

> *Dodinctun* 1086 DB
> *Dodintone, Dodynton* 1086 DB, 1167 P, 1209 *For*, 1281 FF, 1318 Cl
> *Dodintun'* 1158 P
> *Dudinton* 1220 Fees 333, 1227 *Ass*
> *Dudington, Dudyngton* 1227 *Ass*, 1228 Bract, 1248 FF, 1253 FF, *For*, 1286 FF, 1293 Cl, 1310 FF, 1325 Cl, 1340 FF, c. 1350 Pat, 1497 FF
> *Dodington, Dodynton* 1252 FF, 1253 *For*, 1270, 1286 *Ass*, 1301 BM, 1314 Cl, 1319 Ipm, FF, 1327 *SR*, 1375 Cl, Fine, 1470 IpmR, 1526 LS, 1551 FF
> *Duddington* 1260 *Ass*, 1598 Cai
> *Doditone* 1260 *Ass*, 1316 FA
> *Dedingtone* 1286 *Ass*, 1535 VE, 1589 FF
> *Dydyngton, Didington* 1442, 1572 FF
> *Doddington* 1766 J

'Dudda's farm' *v.* ingtun, with the same curious development of *i* from *u* that we have in Denham, Dinton (Bk, W), Dinnington (Nb) and in Dinton (Nth), DB *Dodintone*. Skeat suggests that the name may have been influenced by the not very distant Dillington, but in view of the parallels quoted it is evident that this development might arise independently.

DIDDINGTON WOOD

boscus de Dodingtone 1255 *For*

Eynesbury

EYNESBURY 84 C 12

Eanulfesbirig c. 1000 Saints
Einuluesberie 1086 DB
Einulfesbiri c. 1125 WMP
Ainesbiri 1163 P
Enolfesburia 12th Ord
Eynebir 1227 *Ass*
Eynesbyr, -bir 1234 FF, 1235 Cl, 1286 *Ass*, 1303 FA, 1327 SR
Eynisbyr, -biry 1248 FF, 1275 Cl, 1313 *Ass*
Eynesbury 1316 FA, 1350 Ipm, 1376 Cl, 1428 FA, 1504 Ipm
Eynysbury 1498 Ipm
Aynsbury 1509 BM
Eymesbury 1577 FF
Ensberry Eliz ChancP
Aisbury c. 1600 *Linc*

This is a remarkable but perfectly clear example of the confusion of two like-sounding names. Originally the burh took its name from a man bearing the common OE name *Ēanwulf*. The diphthong in this name would by the time of the Conquest have become *æ*. Then this name became difficult to distinguish from late OE *Einulf* found in the form *Aegenulf, Agenulf*, as the name of a moneyer of Ethelred the Unready and as *Ainulf(us)*, *Einulf(us)* in various late OE documents. The name is continental rather than English in origin (*v.* Forssner s.n. *Aginulfus*). The substitution of such for an English name may have been assisted by the fact that after the founding of the priory in the 11th cent. Eynesbury became a centre of foreign influence.

Godmanchester

GODMANCHESTER 74 H/J 13/14

Godmundcestre 1086 DB, 1173 P
Gutmuncetre 1146–54 BM
Gudmencestre c. 1150 *HarlCh* 83 B 6

Gudmundcestria 1168 P, 1177 P, 1286 Ch
Gum(m)uncestre 1175, 1177 P
Gumencestre 1189 ChR
Guncestre 1189 ChR, 1362 Cl, 1526 LS
Gumcestria 1194 BM
Gumecestre 1197–8 *P*, 1217 Pat, 1219 FF, 1227 *Ass*, 1236 Ch,
   1245 *For*, 1327 *SR*, 1361, 1381 Cl, 1392 Ch, 1467 BM
Gommecestre 1267 Ch
Gomecestria 1285 BM, 1334 Ipm
Gummecestre 13th AD iii, 1308 Cl, 1324 Fine
Gurmund(es)cestre 1302 Orig, 1305 Cl
Gormecestre 1316 FA
Gormancestre 1322 Inq aqd
Gomecestre 1334 Ipm
Gunnecestre 1353 Orig
Gurmecestre 1361 Fine
Godmechestre 1380 Cl
Gurminchestre 1381 Cl
Gumchestre c. 1460 *Linc* (ter)
Gurmencestre 1485 FA
Gumcestre, Gumestre, Godmonchestre 1513 LP
Gumecestur 1521 FF
Gumycestre 1529 FF
Godmanchester 1535 VE
Godmanchester al. *Gunecestre* 1597 FF

The 'chester' (*v.* ceaster) recorded here is the Roman station
south of Ouse, generally and probably, though not quite certainly, identified with the *Durolipons* of the Antonine Itinerary.
The first element in the name is either *Guðmund* or *Godmund*, the latter found also in Goodmanham (Y) and Gumley
(Lei). In each of these names, as in Godmanchester, later forms
occur with a spelling *Guth-* or *Gut-*, and in Gumley and Goodmanham, if derivation from *Godmund* were not proved by early
OE spellings, it would be uncertain whether *Godmund* or
Guðmund lay behind the medieval forms. The best explanation
of these *Guth-*, *Gum-* forms is that the change of *o* to *u* results
from anticipation of the *u* of the second element of the compound *Godmund*.

It is clear that in the *Gurm-* forms we have the results of a pseudo-historical tradition. William of Malmesbury in his *Gesta Regum* (ch. 121) says that the Danish king Guthrum (Alfred's adversary) was called in English *Gurmundus*. From the resemblance of *Gurmund* to *Gudmund* a tradition that Guthrum founded Godmanchester must have arisen at least as early as the 14th cent., though the first reference to it is in Camden's *Britannia* 384 (1594 ed.), 'sed antiquito hoc sub Saxonibus nomine, a Gormone Dano Gormoncester vocari cepit.'

LATTENBURY HILL

Lodona beorg 1012 (12th) Proc Soc Ant (New Series) iii. 49

Professor Ekwall has communicated the following note upon the name:

The OE *Lodona* looks like a gen. pl. and it might be a tribal or folk-name. But as in this charter OE *lacu* appears twice in the dat. as *laca*, it is admissible to take *Lodona* to be the gen. of a *Lodon* f., which would normally have been *Lodone*. In either case it is tempting to compare the name with the river-name Loddon (Berks), Loddon (Nf) (probably an original river-name too), and Lothian (Sc). All no doubt belong to Brit **lutā*, 'mud.' Early forms of Lothian show a good deal of resemblance to *Lodona*, as *Lodoneum* 1098, *Lodonia* 1127 (see Förster, *Engl St* lvi 228 ff.). It is hardly worth while speculating upon the exact British ground-form of *Lodona*, if this is correct. Presumably the district round Huntingdon had a British name derived from **lutā*, 'mud,' and the name meant 'the fen-country.' Such a name would suit the locality very well indeed. We may thus perhaps assume that *Lodon* was an OE name of the Hunts district, a name adopted from their Celtic predecessors, and that Lattenbury is 'the hill in the Lodon district.' If, on the other hand, *Lodona* is a gen. pl., it may belong to a folk-name meaning 'the Lodon people.'

RAVENSHOE (lost)

Rauenesho Steph (1286) Ch, Hy 2 (c. 1230) *Warden* 22, 1279 RH

'Raven's hoh,' the first element being a name of either English or Scandinavian origin.

Great Gransden

GRANSDEN 85 D 1

Grantesden(e) 1086 DB *et passim* to 1485 FA, *(Mekel)* 1339 Pat
Grantendene 1168 P (p), Hy 2 (Hy 3) *St Neot* 90, Hy 2 *HarlCh*
  83 B 41
Grantesdon 1199 Cur, 1315 Ipm, 1377 Cl
Grancenden 1200 FF
Grantedone 1227 *Ass*
Grauncenden 1245 FF
Grauntedene 1276 RH
Magna Granteden E 1 BM
Grauntesden 1289 BM, 1296 Ch, 1314 Cl, 1382, 1426 IpmR
Grantisden 1315 Cl, Ipm
Grandesden Hy 4 AD i, 1422 IpmR, 1526 LS
Grayntesden 1428 FA
Graundesden 1535 FF
Grennesden 1536 FF
Gransden Berristeed 1598 FF

There is mention in the Croyland Cartulary (89 *b*) of a man
named *Grante*, and it would seem that here we have the same
pers. name, hence 'Grante's valley' (*v.* denu). The 1598 form
is for 'bury-stead,' i.e. manor-site.

HARDWICKE

Herdewik 1227 *Ass*
Grantesden Herdwyk 1306 FF
v. heordewic.

LEYCOURT [legət]

Lecote 1227 *Ass*, 1260 *Ass* (p)
Leycote cum Herdwyke 1286 *Ass*
Leycote 1539, 1544 FF
'Cottage(s) by the clearing' *v.* leah, cot.

Eynesbury Hardwicke

HARDWICKE

Herdwich 1209 Abbr
Herdwic 1234 FF

Herthwik 1260 *Ass*
Herdewik 1279 RH
Puttokesherdwyke 1286 *Ass* (p), 1334 FF
Puttukes Herdewik 1309 BM
Hardwyk 1318 Ipm
Pottokesherdewyk 1319 Ipm
Putteshardwyk 1517 FF
Puddock Hardwick 1766 J, 1826 G

v. heordewic. This seems to have been at one time distinguished from the numerous other Hardwicks by its owner's name of *Puttoc*, which is on record in OE as the second name of Ælfric *Puttoc*, archbishop of York, and was probably by origin a nickname from *puttoc*, 'kite,' though that word is not actually on record until the 15th cent. in the form *puttock*. It may, however, represent a diminutive form of the well-recorded OE pers. name *Putta*.

CALDECOTE

Caldecote 1242 FF
Caudicot 1260 *Ass*

v. ceald, cot.

LANSBURY FARM

Launceleynsbury 1506 *FF*

An example of the manorial use of burh, 'manor of the Launcelyn family.' The *Launcelyn* family were here in 1160, cf. St Neots Cartulary f. 38.

WEALD

Weld(e) 1219 *For*, 1270 *Ass*, 1279 RH, 1286 *Ass, For*, 1310, 1327 FF, 1410 BM, 1443 IpmR
Wald(e) 1220 FF, 1270 *Ass*
Waude 1227 *Ass*
Wolde 1286 *Ass*
Welde juxta S. Neotum 1287 FF
Weelde 1558 FF

v. weald, with much fluctuation between the two dialectal forms *weald* and *wold* which develop from it.

Hemingford Abbots and Grey

HEMINGFORD 75 H 1/2

Hemminggeford c. 1000 Hist El

Hemmingaford 1012 (12th) Proc Soc Antiq iii. 49

Emingeford, alia Emingeford 1086 DB

Hamicheford c. 1150 BM

Hemingeford 1150 BM, 1220 Fees 333 (*Trublevill*), 1236 Cl, 1272 FF

Hemmingeford, Hemmyngeford 1164 P, 1220 FF, 1235 Cl, 1236 FF, 1242 Fees 921, 1286 *Ass* (*iuxta Huntedon*)

Emmingeford 1202 *P*

Hemmingford, Hemmyngford 1218 FF, 1276 RH (*Abbatis*), 1289 Cl, 1307 BM (*iuxta S. Ivonis*), 1316 FA (*Grey*)

Hemingford, Hemyngford 1248 FF (*West*), 1252 (c. 1350) Rams (*Magna Parva*)

The family name *Turbervill* is found in that form in 1272 FF, 1324 Cl, Ipm, as *Trubbeuill* al. *Turbeuill* in 1279 *Ass*, as *Tribeluill* in 1281 FF, *Trubuill* and *Turblevill* in E 2 Orig, *Turbeuill* in 1315 FF, *Trumbevill* in 1315 Cl, *Trubbevill* 1316 Fine, *Turbduill* 1337 FF

'Ford of the people of Hemma or Hemmi.' Both names are recorded in LVD, and in ultimate origin are probably short forms of some compound name in *Hǣm-*, such as *Hǣmgils* (cf. *Bledda* from *Blǣd-*). It may be noted that the name Hemingford is recorded too early for derivation from ON *Hemmingr* to be likely. Cf. *Hemmingebroke* in Yaxley (*Thorney* 7 a). *Abbots* from a holding of the Abbot of Ramsey (DB), also known as *West* Hemingford and Hemingford *by Huntingdon*. Hemingford Grey was so called from the Grey family who had a holding here in 1276 (RH), and was also known as Hemingford *by St Ives*. The *Turberville* family, who also had a holding in Hemingford, took their name in all probability from Trubleville in Normandy. Cf. Fabricius, *Danske Minder in Normandiet*, 268. Abbots Hemingford was in Great Hemingford, the Grey holding was in Little Hemingford (*VCH*).

LITTLEBURY

Littlebire, Litlebyri 1209 *For* (p), 1327 *SR* (p)

Lytle Biryhill 13th AD iii

The name is self-explanatory except that we have no means of knowing just what the *bury* here may be. *v.* burh.

Hilton

HILTON 85 A 1

Hiltone 1196 FF (P) *et passim*
Hulton 1227 *Ass* (p)

'Hill-farm' *v.* hyll, tun. There is no actual hill, but the village is slightly above the level of the surrounding country.

Huntingdon

HUNTINGDON 74 H 14

Huntandun 973 (c. 1300–25) *Thorney* 1 *b*, c. 950 (921 A) ASC,
　c. 1000 *Cragg*
Huntadun c. 1000 *Cragg*
Huntedun 1086 DB
Huntedon 1199 FF, 1230 Cl
Huntendonia, id est Mons venatorum 12th HH
Huntendon 1212 FF
Huntindon 1225 Pat, 1227 Ch, 1259 FF
Huntyngdon 1286 *Ass*

The OE *huntan-dūn* can at once be translated 'hunter's hill,' but such a derivation though possible is not on the whole very probable. In Huntingford (Wo), where association with hunting is certain, the original name was *huntena ford*, 'ford of the hunters,' in the plural. It is more likely that the well-recorded OE pers. name *Hunta*, which survived into the 12th cent., is contained here. It seems also to occur in Huntham (Sa) and Huntington (Y). This pers. name may have meant 'hunter,' or equally well may be a short form with a *t*- suffix of one of the numerous pers. names beginning or ending with *Hūn*. The latter is made probable by the existence of a mutated form **Hyntel*, **Hyntil*, preserved in Hintlesham (Sf). The dun in Huntingdon must be the stretch of rising ground between the Ouse and the valley in which King's and Abbot's Ripton lie.

HINCHINGBROOKE

Hychelingbrok 1260 *Ass*
Inchinbrok 1378 BM

U

Fynchyngbroke 1402, 1412 BM
Fynchynkbroke 1478 BM
Fyncheynbroke 1478 BM
Hynchenbrok 1526 LS
Hynchingbroke 1535 VE

This is a difficult name. Change of initial *h* to *f* is, however, not without parallel (*v.* Filbert Haugh in PN NbDu 85). In any case we may assume that *h* is the original initial sound. An OE *Hyncelingabroc* would explain the forms here given. There is on record an OE *Hynca* (LVD) from which an *-il* diminutive, *Hyncel* could easily be derived. Hence we may interpret the name as 'the stream of Hyncel's people.' Cf. Billing Brook *supra* 193 for a stream-name of this type.

Midloe

MIDLOE 84 A 11

Middelho, Myd- 1135–60 (c. 1350) Rams, 1252 Ch, 1539 FF
Midelho c. 1200 (c. 1230) *Warden* 21 *b*
Middellowe 1535 VE
Medlowe 1567, 1594 FF

'Middle hill' (*v.* hoh), possibly in relation to Southoe or, still more probably, to the lost *Westhoe* (*v.* Southoe *infra* 266) on the one hand and the lost *Mulsoe* on the other.

MULSOE (lost)

Molesho 1257 BM
Mulsho 1540 FF
Mulso(w)e 1567 FF, Eliz ChancP

'Mul's hoh,' cf. Molesworth *supra* 246.

Offord Cluny and Offord Darcy

OFFORD 74 J 13, 84 A 13

Vpeforde, Opeforde 1086 DB
Uppeford 1195 BM, 1199 Cur
Opford 1199 Abbr, Cur, 1202 Cur, 1205 *P*
Oppeford 1199 Abbr, 1202 Cur
Upford 1199 Cur, 1230 Bract

Offord 1200 *Fine*, 1220 Fees 333 (*Willelmi Daci*), 1257 FF
(*Clunye*), 1260 *Ass* (*Daneys*), 1261 FF, 1270 *Ass*, 1271 FF,
1286 *Ass*, FF, 1295 FF (*Deneys*), 1303 FA (*Daneys*), 1322
Ipm (*Deneys* al. *Daneys*), 1428 FA (*Deneys*), 1474 FF (*Dacy*),
1525 FF (*Dacy* al. *Danes*), 1596 FF (*Darcye* al. *Dacie*)
Hupford, Uppord 1200 FF
Ufford 1260 *Ass*, 1261 FF, 1327 *Fine*, 1373 IpmR, 1392 BM
(*Denys*), 1399 IpmR, 1435 IpmR (*Deneys*), 1439 FF (*Deynys*)

Clearly 'up(per) ford,' but in relation to what other ford is
not certain. It may however be noted that the Ouse is crossed
at Offord Cluny for the first time above Huntingdon. Offord
Cluny[1] from the monks of Cluny who held land here in DB.
The *Dacy* family held land here in 1220 (Fees 333) and the
widow of Joh. le *Daneys* in 1242 (ib. 923).

COTTON

Cotes 1086 DB

OE *cotum* (dat. pl.), 'cottages,' *v.* cot. DB has the nom. pl.
form.

THORN (6″)

Thorn' 1279 RH

Evidently so called from some prominent thorn-bush.

Great and Little Paxton

PAXTON 84 B 12

Parchestune 1086 DB
Pachstone 1086 DB

---

[1] Mr Candlin of Offord Cluny has kindly furnished a full list of modern
field-names from that parish with sketch-map. Unfortunately no early
forms of these names have been found so that it is impossible to do much
with them. We have a *Ray* field close to the Ouse, clearly taking its name
from that river (*v.* ea). There is also a *Rowboro* or 'rough hill' and a Parlour,
earlier *Parlow*, field. This is probably from *pere-low*, 'pear-tree hill,' and a
good example of the senseless corruptions which field-names often undergo.
More interesting is *Innage* ground. The meaning is unknown, but it looks as
if it were the same field-name found in Dunstable as *Les Westinnynges* (1371),
les North Innynges with the same development of final *-age* for *-inge* which
we have in Lymage and *Pillage* for Pillinge, *v.* pp. 87, 270. Presumably this
is the same as the word *inning*, denoting first the reclamation of waste or
marsh land and then the land itself. If so the history of this word is carried
back some two hundred years earlier than the first example in the NED.

Pacstonia Hy 2 *D and C Linc* D ii 90/3 no. 23, 1164 BM
  (*Magna*)
Paxton ib. nos. 20, 21, 24, 25, 1227 *Ass et passim*
Pastona 1176 BM
Parua Paxton 1227 *Ass*
Magna Praxton 1245 *For*
Paxtone Parva 1316 FA
Magna Paxton al. *Muche Paxton* 1588 FF

The best explanation that can be offered of this name is that it was originally *pearroces-tun*, 'farm of or by the enclosure.' Cf. *parkesriding* (*Newn* 116 *b*). It is possible however that the reference may be to a fishery in the Ouse, cf. Parkhall *supra* 223. This would explain the DB form and (with metathesis of the *r*) the 1245 one. Early loss of *r* might be accounted for by the strangeness of the type of name or by the existence of such a name as *Pæcc*, for which *v.* Skeat, PN Hu 345. On the other hand, the Evesham Cartulary (MS of *circa* 1195) records a pers. name *Perruc* which is not found in OE records, and indeed is not easily explained, but may possibly be sufficiently ancient to enter into the present name. In any case the two forms with *r* in the first element should not be ignored in any attempt at the explanation of Paxton. *v.* Addenda.

GALLOW BROOK

Galuwe 1286 *Ass* (p)

Probably from a place where a gallows was erected.

HAIL BRIDGE

Hailebruge 1265 Coroner, (*juxta Haileruding*) *St Neot* 70 b
Halyisbrigg 1276 RH
Haillebrugg 1377 IpmR

The bridge over the Kym, once known as the Hail River, *v. supra* 7. For *ruding v.* hryding.

MEAGRE (Fm) [megri]

Maugrey 1284 FF
Maugrith 1376 Cl
Mawgrych 1451 IpmR
Mangryth (sic) 1452 IpmR

Gt and Little Maugrey 1567 FF
Magrey 1571 FF
Meggary Farm 1836 O

This is probably from *Mæðelgār-rīð*, 'Mathelgar stream,' *v.* rīð, and cf. Sawtry *supra* 195. For the earlier stages of the development of the name, cf. Maugersbury (Gl) from *mæðelgares byrig* (BCS 882).

St Neots

ST NEOTS [sənt niˑts], [snouts], [sniˑdz] 84 C 12

S' Neod 1132 E (12th) ASC
villa S. Neoti 1203 FF
Saint Nede 1310 Cl
St Nyot's 1329 Ipm
Seinthede 1381 Cl
Seynt Nedys 1513 FF
Mon. S. Neoti vulgo nuncupatur Sainte Need 1542 BM
St Edes 1558 BM
St Noots 1576 Cai
St Neotes al. *Saynt Nedes* 1583 FF
St Neots vulgo *St Nedes* c. 1750 Bowen

v. Addenda.

MONKS HARDWICK

Herdwic, Herdewyk Monachorum 1203, 1313 FF
Hungry Hardwick 1766 J

v. heordewic. This Hardwick was distinguished from others by being known as *Monks'* from its possession by the monks of St Neots. It also seems to have had a reproachful epithet 'Hungry.'

WINTRINGHAM

Wintringeham Hy 2 (Hy 3) *St Neot* 46 *b*
Wintringham, Wynt- 1227, 1270, 1286 *Ass et passim*
Winteringham 1270 *Ass*

'Homestead of the people of *Wintra* (actually on record) or *Winter* (which is suggested by such a p.n. as Winterslow (W)).' Cf. Winteringham (L), Wintringham (Y) and Winterton (L).

Southoe

Southoe 84 A 12

Sutham 1086 DB
Sudho 1187 P, 1220 Fees 333
Suho 1220 FF *et saepissime*
Southo(u) 1255 FF, 1274 Ipm, (*Wynchestr'*, *Lovetoft*) 1276 RH
Sutho cum Westho 1286 *Ass*

The meaning is obvious for there is a well-marked hill here (*v.* hoh). It was 'south' in relation to a lost *West*-oe, which may have been the hill just to the north of Little Paxton Wood. This place was the *caput* of the Feudal Honour of *Southoe Luvetot*, held by Nigel de Luvetot in 1166 (RBE 372, 3). *Luvetot* is a Norman p.n. with Scand. *tot* or *toft* as the second element. It is once found as *Ulvtofte*. It is difficult to say which of the two is the original and which the translation (cf. Fabricius, *Danske Minder*, 314). In 1242 (*Fees* 922) the Earl of Winchester held the third part of a knight's fee in Southoe.

Boughton (Lodge)

Buchetone 1086 DB
Bouton 1199 Cur, 1225, 1248 FF, 1303 *SR*, Cl, FA, Ipm, 1316 FA, 1317 Cl, 1318, 1350 Ipm
Buweton 1220 Fees 333
Bugheton 1225 FF
Bukton, Bucton 1340 FF, 1351 Ipm
Boudon 1360 FF, 1363 Ipm
Bowton 1375 Cl
Bowedon 1383 FF

OE *Bugan-tūn*, 'Buga's farm.' The *Buche-*, *Buwe-* and *Buge-* forms make it impossible to take this, with Skeat, as an example of the common OE *bōc-tūn*, 'beech-tree farm.'

Wray House

Ray Houses 1766 J

The modern form is clearly corrupt. The place lies down by the Ouse and the older form shows that we have one of the common examples of *Ray* developing from OE ea, 'river,' noted under that word in EPN. The modern form shows the influence of the north country dialectal *wray*, 'nook, corner' (*v.* **vra**). It is quite possible that a similar corruption has taken place in Wray Common in Reigate (Sr) and that Wray is really the same word as the *Rei-* in that name, which in its turn seems to be nothing but the word *Rye* or *Rey* from OE **eg** noted in PN Bk 205.

Fen Stanton

FEN STANTON 75 J 2

Stantun 1012 (12th) Proc Soc Antiq iii. 49

Stantone 1086 DB *et passim*

Stantun(e) c. 1150 (c. 1300-25) *Thorney* 168 *b*, 1234 Cl

Staunton 1227 *Ass*, 1229 Bract, 1234, 1236, 1238 Cl, 1253 Ch, 1254, 1257 FF, 1286 *Ass*

Stanton Gisebryt de Gante 1245 (c. 1350) Rams

Stanton Grysebryke, Gryseby 1257 FF, 1280 Ch, 1286 *Ass*, 1292 Ch, 1303 FA

Fenstanton 1260 *Ass*

Staunton Gryseby 1272 FF

Fenstaunton 1315 Ch

Fen Stanton al. *Fenny S.* 1323 Inq aqd

Fenistanton, Fenny- 1344 FF, 1353 Cl, 1354 Ipm, 1372, 1377 Cl, 1526 LS

Fennystaunton Hy 6 IpmR, 1489 FF, Eliz ChancP

Fennystanton 1836 O

For this common but somewhat ambiguous p.n. *v.* **stan, tun**. *Fenny* or *Fen*, probably in contrast to Long Stanton (C) which is on rather higher ground. Cf. similarly Fen and Dry Drayton (C). The manorial addition is due to a holding of *Gislbert* or *Gilbert de Gaunt* who, in 1228, held land in Stanton and was followed there by two other Gilberts of Gaunt (*VCH*).

LAKE BROOK (6″)

(*oð ða ealdan*) *laca* 1012 (12th) Proc Soc Antiq iii. 49

v. **lacu**, the reference being to some slow-moving fen-land stream.

Great Staughton

STAUGHTON 84 A 10

Stoctun c. 974 (c. 1250) BCS 1306

Tochestone 1086 DB

Stottun 1163 P

Stocton(a) 1163 P (Chanc Roll), 1209 FF, 1232 FF (*Magna*) et *passim*

Stoctun c. 1198 BM, 1318 FF

Stoucton 1260 *Ass*

Stokton 1287 *Ass*, 1316 FA, 1378 Cl, 1502 Ipm

Stoghton 1358 Cl, 1366 FF

Stoughton 1504 Ipm

Moche Staughton 1566 BM

Stowghton Magna al. *Stocton* 1592 FF

'Farm-enclosure made of stumps or stocks' *v.* stocc.

AGDEN (Green)

Accadena 1124–8 (14th) Scottish Hist. Review xiv. 371 (*bis*)

Hakeden 1227 *Ass* (*boscus de*)

Akeden 1241 FF, 1248 *For*, 1327 *SR* (p)

Agden Grene 1553 FF

The first form makes derivation from an OE *āc-denu* (*v.* ac, denu) highly improbable and suggests that the first element was the ancient OE pers. name *Acca*. In the 12th cent. an Anglo-Scandinavian pers. name *Acca* (or *Akke*) existed as a short form of the common ON *Ásketill*. It is however improbable that this name would appear in combination with the OE *denu* in a Huntingdonshire p.n.

BEACHAMPSTEAD (lost)

Bechansted 1141–53 BM

Bich(ch)amestede, Bych- Hy 2 BM, 1227 *Ass*

Bi(c)ch(h)amstede, Bych- 1163 P, 1197 P, 1231 Bract, 1248, 1316 FF, 1318, 1358 Ipm, 1418 FF

Bicchamestud 1228 FF

Bichamstude 1245 *For*

Bichehamstede, Bych- 1248 FF, 1429 AD i

Bechamstede 1341, 1478, 1545 FF

Bichampstede, By- 1377 FF, 1378 Cl, 1489 Ipm

Beauchampstede 1484 FF

Beachamstede 1577 FF

'Bicca's hamstede.' For the pers. name and for the irrational lengthening of the vowel we may compare the history of Beachendon (Bk). Desire to dissociate the name from *bitch* and ready association with *beech* have both done their work in these names as in Beechburn (Du), locally *Bitchburn.*

BLASWORTH (lost)

Blaywurth 1227 *Ass*
Blaysworth, Blayswurth 1241 FF, 1254 Pat, 1279 RH, 1286 *Ass*
Blaesworth 1255 *For* (p), 1260 *Ass*
Bleyswrth 1260 *Ass*
Bleieswrth 1279 RH (p)
Blesworthe 1279 RH, 1286 *Ass*
Blayesworth 1286 *Ass*
Blaythesworth 1286 *Ass*
Blysworth 1286 *Ass*
Blasworth 1535 VE

No certainty can be attained with regard to this name. With the second element worð and a first element in the possessive, the presumption is that the former is a pers. name. The form *Blaythesworth* found in 1286, which at first sight seems to provide a clue to a fuller form, is probably to be taken rather as a bad spelling, due to anticipation of the *th* which is to come later in the word. The diphthong *ay, ey* has arisen presumably from the loss of an intervening consonant, which might be *g* (when the loss would be perfectly normal) or *th* (when we should have an example of the common intervocalic loss of *th* due to AFr influence, cf. Sawtry, Meagre *supra* 195, 264). There is an OE *blæge,* 'gudgeon, bleak,' which may conceivably have been used as a pers. name by way of nickname, cf. the common metaphorical use of *gudgeon* of 'a greedy person.' This would readily explain the later forms. Alternatively OE *blēað,* ME *blethe,* 'soft, effeminate,' may equally well have given rise to a nickname which would, though not quite so readily, explain the later forms.

DILLINGTON

Dilingtun 674 (12th) BCS 32, c. 1100 (c. 1350) Rams
Dellinctune 1086 DB
Dylynton, Dili-, Dily- 1241 FF, 1253 BM

Dilington, Dyl- 1245 *For*, 1297 Ipm, 1303 Ch, FA, 1310, 1342
   Cl, 1358 Fine, 1359 Ipm, 1365 Cl, 1464 BM
Dulintone 1255 *For*
Delington, Dely- 1286 *Ass*, 1372 BM, 1377 FF, 1378 Cl
Dillyngton, Dyll- 1323 Ipm, Cl, 1340 Ipm, 1428 FA
Dellington al. *Dyllington* 1599 FF

In the forged charter BCS 32 there is mention of a *diling-broc* among the boundaries of Dillington. The probability is, not that the hamlet takes its name from the stream, but that both alike take their name from the same person, a fairly common phenomenon in place-names (cf. the history of Waddesdon and Beachendon, PN Bk 138–9). The later forms point clearly to an OE *Dy(l)la* as first element, probably a derivative of the adj. *dol*, 'foolish,' and allied to the pers. name which lies behind Dullingham (C), cf. Ekwall PN in *-ing* 139. Hence 'Dyl(l)a's farm and brook,' *v.* ingtun.

GAYNES (Hall)

   Gaynes 1389 Cl
   Gaynes Hall 1593 FF

This is a manorial name from the Engayne family, who had land in Great Staughton (cf. FF 1288). William *Ingania*, a member of this family, held a manor in Gidding in 1086. On the family see Round in VCH, Northants, i. 294.

LYMAGE (Fm)

   Limminges, Lymmynges 1209 *For*, 1237 Cl
   Lymmynge, Limmynge 1241 FF, 1279 RH, 1302 Ch
   Limmig' 1242 P
   Limming' 1255 *For*[1]
   Liminges 1286 *For*
   Lemynge 1350 Ipm, 1354 Orig
   Lymenge 1354 Orig, 1358 Fine, 1389 IpmR
   Lymyng 1375 Cl
   Lymmyng al. *Lymmage* 1544 FF

---

[1] The reading as printed in *Select Pleas of the Forest* (Selden Soc. xiii) is *Limininge*. It is true that in this case we have simply minims with no marks upon them at all, but when the name is repeated afterwards there are faint dots to mark the *i* on either side of the six minim strokes.

This name is dealt with by Ekwall in his PN in *-ing*, and he definitely disassociates it from Lyminge (K), earlier *Limingae*, *Liminiaeae* (BCS 97, 160). He is of course right so far as the suffix is concerned. In one case we have *Limm + inge(s)* and in the other *Limin + ge*. The first element in the latter name he takes to be the well-established river-name *Limen* found also in Leam (Wa) and Lemon (D). The first element of the former he takes to be a river-name *hlymme*, 'torrent,' connected with OE *hlymman*, 'to roar.' Such a name does not seem very appropriate for a Huntingdonshire stream and it may be that after all the two names are identical. If the earlier and fuller form of the Hunts name was *Liminingas* this would readily become *Limminges* by assimilation. Such an assimilation of *mn* to *m(m)* is common as in *stemn* becoming *stem* and *emn* becoming *em*, *remn* or *ramn* becoming *rem* or *ram* in the history of OE *stefn*, *efn* and *hræfn*. No certainty is possible, but it is clear that Lymage must take its name from the small stream on which it stands, whether its name was *Hlymme* or *Limen*, and that the suffix is -ingas denoting the dwellers on that stream. For the development of the *-age*, cf. Pillinge *supra* 87.

WEST PERRY

> *Pirie* 1086 DB, 1227 *Ass*, 1286 *Ass* (*Louetot*) *et passim*
> *Perihe* 1219 FF
> *Pery* 1463 FF
> *Pury* 1478 FF
> *Pyrry* 1484 FF

'Pear-tree' *v.* pirige.

PERRY WEST WOOD

> *boscus de Pirie* 1286 *For*

STAUGHTON BRIDGE

> *atte Brigge* 1325 Abbr (p)

STAUGHTON MOOR

> *Mora de Stocton* 1227 *Ass* (p)
> *Mora* 1279 RH

Tetworth

TETWORTH 84 E 13

Tethewurða c. 1150 BM
Tetteworth 1209 For (*et passim*)
Tetteswrthe 1244 (c. 1350) Rams, 1286 QW

'Tetta's enclosure' *v.* worð. For personal names *Tetta, Teota*, and *Teotta*, we may compare Teddington (Wo), *Teottingtun* BCS 236, Tettenhall (St), ASC *s.a.* 909 D *Teotanheale*, Tedburn (D), *Tettanburna* BCS 1331.

Toseland

TOSELAND 84 B 13

Toleslund 1086 DB[1], c. 1180 *D and C Linc* D ii 90/3 no. 24, 1255 *For*
Touleslund 1231 FF, 1286 QW
Tholeslund 1241 FF
T(h)ouleslound 1281 FF, 1308 Ipm
Toulislond, Toules- 1284 Cl, 1303, 1316 FA, 1317 Ipm, 1319 Pat, 1329 FF, 1346 Orig, 1361 Cl, 1362 IpmR
Tollesland 1321 Orig
Toulesland 1324 Ipm
Tow(e)slond 1364 FF, 1428 FA, 1443 IpmR
Touslond 1378 Cl
Touseland 1396 IpmR
Towesland 1507 FF
Towseland 1540 BM, 1568 FF

The second element in this name is the ON lundr, probably with the heathen religious associations attached to it, cf. Holland *supra* 220, 252. It is possible with Skeat to identify the first element with the Danish earl called *Toglos* in the Anglo-Saxon Chronicle, and the *Toli* of the *Liber Eliensis*, an earl of this district, who fell at the battle of Tempsford in 921. It should be added that while the identification of *Toglos* and *Toli* seems certain, it is impossible to bring these two names into elation with one another. *Tóli* is found, though rarely, in ONorw but is more common in ODan *Toli* and in OSw *Tole*.

---

[1] The vill is not mentioned in DB, this is the form of the hundred-name derived from it.

Toglos is unknown, though Björkman suggests (NP 142) that it might be a nickname from ON *tauglauss*, 'ropeless,' applied perhaps to a man who at a critical time had not got the right ropes on his ship. If that was his real name the writer of the *Liber Eliensis* being unfamiliar with this name may have turned it into a more familiar one or (just conceivably) the earl may have had an alliterative nickname and been called *Toli Tauglauss*. Whatever be the true relation of *Toglos* and *Toli* there is a good deal to be said for thinking that both names lie behind Toseland. *Toglos* would readily explain the persistent ME *Toul-*, otherwise very difficult to account for. *Toli* on the other hand is needed for early *Tol-* forms. The reduction of the suffix *-los* has its parallel in Brocklesby (L) which appears as *Brochelesbi* in DB but which (from other early forms) certainly contained ON *Bróklauss*, a nickname meaning 'breechless.' So also Scamblesby (L) from ON *Skamlauss*, 'shameless.' On the other hand, the gen. sg. of ON *Tóli* is *Tóla*, and we should have expected an *s*-less form as in Tolthorpe (R) if that name had lain behind Toseland.

TOSELAND WOOD (6")

boscus de Tolleslond 1245 *For*

Waresley

WARESLEY [weizli] 84 D 14

Wedreslei(e), *Wederesle* 1086 DB

Weresle(a), *Wereslai* c. 1130 (Hy 3) *St Neot* 90, 1169 P, 1193–8 BM, 1200 FF, 1220 Fees 333, 1227, 1291 *Ass*, 1289 Cl, 1303, 1316 FA, 1323 Ipm, 1330, 1378, 1389 Cl, 1485 FA, 1533 FA, 1535 VE

Waresle(g) 1199 Cur, 1273 Ipm, 1286 FF, 1309 Orig, 1316 FA

Werisleg 1224 FF

Wersle(y) 1299 BM, 1351 FF

Werisley, *Werysle(y)* 1323 Cl, 1327 *SR*, 1340, 1425, 1484, 1549 FF

Wyrseleye al. *Wersleye* 1323 Ipm

Worsle 1377 Cl, 1384 BM

Wyersley c. 1400 *Linc*

Waresl(e)y 1435 AD i, 1496 Ipm
Whearesley c. 1540 *Linc*
Weyrysley 1554 FF
Werisley al. *Werysley* 1571 FF
Warseley 1574 BM
Waresley al. *Warslye* 1592 FF

Almost certainly we should not attach too much weight to the DB form here. There is some evidence for an OE pers. name *Weder* (cf. Skeat on Wetheringsett (Sf) in PN Sf 85), but it is impossible to believe that the *d* could have disappeared so completely or that if it had ever really been in the name the phonological development would have been what it was. The DB form is best regarded as a scribal error, probably due to a misreading of an *æ* in the original returns. In late OE script the *e* of this combination occasionally rises far above the line so that *æ* might by uninterested copyists be read as *ed*. Such a confusion would be made easier by the existence in the neighbouring county of Cambridge of a Hundred-name *Wederlea*, with which the compilers of the Huntingdonshire DB may well have been familiar. If the DB form may be left out of consideration, the name may be regarded as derived from OE *Wæresleage* (v. leah), *Wǣr* being a shortened form of one of the numerous OE pers. names in *Wǣr-*, found also in Waresley (Wo), which is *Wæresleage* in BCS 362, and has later forms just like those of the Hunts p.n. The same name occurs in Warwick, and in Warsop (Nt).

BULBY HILL

Boleby 1286 *Ass* (p)

As the only evidence for this *by*-name is from a pers. name and, apart from this name, it is very doubtful if there are any by-names in Hunts at all, it is probable that the man who is mentioned in the Eyre Roll came from Bulby (L), DB *Bolebi*.

WARESLEY DEAN BROOK (6″)

Weresleedene 1306 Abbr

Self-explanatory.

Hail Weston

HAIL WESTON 84 B 11

Heilweston 1199 Cur
Helweston 1209 *For*
Haileweston, Hayle- 1219, 1248 FF, 1270 *Ass*, 1284 FF, 1286
    Ass, 1376 Orig
Halewestan 1247 Cl
Haylweston, Hail- 1315, 1346 Ipm
Hayle Weston 1427 BM, 1502 Ipm
Heyleweston 1534 FF

'West-farm on the Hail river,' but what it was west in relation to, it is impossible to say, unless we take it to be a settlement
from Eynesbury-St Neots. For the river-name *v.* p. 7. The
river-name is still preserved in Hail Bridge *supra* 265.

BROOK END (Fm) (6″)

atte Broke 1286 *Ass* (p)

Self-explanatory.

RUSHEY FARM

Russho, Rysho 1286 *For*

'Rush-grown slope' *v.* rysc, hoh. The hoh is a very small one.

Yelling

YELLING 85 B 1

G(h)ellinge, Gelinge 1086 DB
Gellinches Steph BM
Gillinge, Gyllinge 1218 FF, 1219 *For*, 1239 FF, 1316 FA,
    1535 VE
Gillinges 1220 Fees 333, 1228 FF
Gilling, Gyll- 1235 Cl, 1253 BM, 1278 FF, 1299 BM, 1308
    Ipm, 1324 Cl, Ipm, 1325 Cl, 1343, 1361 Ipm, 1428, 1485
    FA, 1535 VE
Gillingg 1276 RH, 1303 FA
Gyllyngs 1333 FF
Yillyng 1344, 1385 Cl
Illyng 1507 FF, 1540 BM

Yellyng 1535 VE
Elyng 1535 VE
Gellyng al. *Yellinge* 1545 BM
Yelling al. *Yeldyng* 1569 FF
Yellinge al. *Gillinge* 1594 FF
Yealding al. *Yealing* 1601 FF

One cannot go further with this name than Ekwall's explanation of it (PN in -*ing* 88) as from an OE *Gellingas* (*v*. ingas) a patronymic plural form, from a name **Giella*, as a parallel to which he quotes the OGer p.n. *Gellingin*.

THE ELEMENTS, APART FROM PERSONAL NAMES, FOUND IN BEDFORDSHIRE AND HUNTINGDON-SHIRE PLACE-NAMES

This list confines itself for the most part to elements used in the second part of place-names or in uncompounded place-names. Under each element the examples are arranged in three categories, (*a*) those in which the first element is a significant word and not a pers. name, (*b*) those in which the first element is a pers. name, (*c*) those in which the character of the first element is uncertain. Where no statement is made it may be assumed that the examples belong to type (*a*). Elements which are not dealt with in the *Chief Elements used in English Place-names* are distinguished by an (n) after them.

(*a*) *Bedfordshire*

ac (b) Ruxox.
æcen (n) (a) Eggington.
beorc Little Barford, *Barkditch*.
beorg (a) Brogborough.
beretun Barton-in-the-Clay.
bræc Brache, Breach.
broc Ashbrook, Brook End, Brookend Green, Brook Farm (3), Brookland, Millbrook, The Old Brook, Sharnbrook.
brom Broom.
bruere, ME (n) Brewershill (?)
brycg Stanbridge.
burh (a) *Aldermanbury*, Kingsbury, Limbury, Medbury, *Sudbury*, Willbury, (b) *Cadbury*, (c) Upbury. See further p. 291.
burna Bourne End, Husborne Crawley, Melchbourne, Woburn, Wootton Bourne End.
byge (n) Beeston.
camp (b) Hanscombe.
cealc Chalgrave.
ceorl Chalton.
clif (b) Hockliffe.
clopp (n) Clapham, Clophill.
cnoll The Knoll.
conynger, ME (n) Conger Hill.
cot(e) (a) Calcutt, Caldecote, Cotton End, Eastcotts, Fancott, Holcot, Lambcourt End, Thorncote, Utcoate, (c) Biscot.
croft (a) *Woodcroft*, Woodcroft, (b) Francroft.
cumb (a) Coombe Farm, *Combe Park*, (b) Pascombe.

x

dal (a) Dallow (?), Whitsundoles.

denu (a) Dane End, Beeston Dean, Dean, The Dean, Honeydon, Pegsdon, Stodden, Yelden, (b) Ravensden, Colesden, Wilden, (c) Stagsden.

dic *Barkditch*.

dile (n) *Dilwick*.

dun (a) Denel End, The Downs, Eggington, Harrowden, Maulden, Pegsdon, Stondon, Warden, Warden Hill, (b) Battlesden, Billington, Cadington, *Elvedon*, Harlington, *Shirdon*, Shillington, Sundon, Toddington, Wensdon, (c) Roxton.

*dyfel (n) Duloe (?).

ealdormann (n) *Aldermanbury*.

ecg Brownage Wold.

edisc Farndish.

eg (a) Ion, Sandy, Turvey, (b) Arlesey, *Fotsey*, Lewsey.

ende (a) Bourne End, Brook End, East End, Hall End, Lower End, Northend, Upton End, Wick End, Wood End (2), Woodend, Woodend Lane, Woodman End, Wootton Bourne End.

fald Faldo.

feld (a) Cranfield, Fielden, Fielding, Greenfield, Northfield, Westfield, (c) Wingfield.

fleot Flitt, Flitton, Flitwick.

ford (a) Barford (2), Eatonford, Girtford, Langford, Salford, Shefford, Stafford, Stanford, *Stapleford*, Stratford, Tempsford, Trevor, (b) Batt's Ford, Bedford.

fox-hol Foxhole.

fyrhþe Marston Thrift, Salem Thrift.

gorstig Goswell End.

græf (a) Chalgrave (?).

grǣfe (a) Chalgrave (?).

grāf(a) (a) Blackgrove, Gravenhurst, Grove, Grovebury, Leagrave, Shortgrove, (b) Collins Grove, Potsgrove.

grange, ME (n) The Grange.

grene The Green.

greot Girtford, *Greathamstead* (?).

hæcc (a) Hatch, Hatch Farm, (c) Sheerhatch Wood.

(ge)hæg (a) Astey, Hay Wood (2), Rowney, (b) Dedmansey, Halsey, Limbersey.

hæð Heath.

haga (a) Haynes (?).

ham (a) Clapham, Higham Gobion, Newnham, Studham, (b) Biddenham, Blunham, Bromham, Felmersham, Pavenham.

hamstede (a) *Greathampstead*, (b) Wilshamstead.

hangra (a) Bramingham, Hanger Wood, (c) Moggerhanger, Polehanger.

har (n) Herne.

heafod *Manshead*, Swineshead.

healh (b) Meppershall, Pertenhall, (c) Renhold.

hearg Harrowden, Harrowick.

hecg Wardhedges.

heordewic Felmersham Hardwick, Kempston Hardwick, Shefford Hardwick.

hid Hyde.

hlaw (a) *Goldenlow*, Henlow, Wenslow, (b) *Bucklow*, *Wadlow*.

hlose Lowe's Wood.

hoh (a) Dallow, Faldo, The Hoo, Hoo (2), Hoo Farm, Hoo Hill, Houghton Conquest, Houghton Regis, How Wood, *Langnoe*, Luton Hoo, Millow, Northey, Salph End, Sharpenhoe, Staploe, Totternhoe, Whitnoe, Yelnow, (b) Arnoe, Backnoe, Bletsoe, *Bolnoe*, Budna, Cainhoe, Dungee, Hipsey, Keysoe, Putnoe, Segenhoe, Silsoe, Winter (Wood), (c) Catsey, Duloe, Gadsey Brook, Galsey, Kidney, Risinghoe.

holm Holme, The Holmes.

holt Holt, Eversholt.

hris Riseley (?), Risinghoe (?).

hrycg (a) Cowridge, Ramridge.

hunger Hungerhill.

hunig Honeydon.

hyll (a) Ampthill, Brickhill, Clayhill, Clophill, Cockle, Denel End, Galley Hill, Hill Farm (4), Hill Ho, Hill Lane, Hungerhill, Oakley Hill, Odell, (b) Picts Hill, Pulloxhill, Roxhill.

hyrst (a) Gravenhurst, (b) Bolnhurst.

ing(as) (b) Knotting, Wootton Pillinge, (c) Kitchen End, Worthy End (?).

ingatun (a) Seddington, (b) Lidlington.

ingtun (b) Cardington, Goldington, Podington, Stevington, Willington (?).

karla-tun Carlton.

lacu Collick, Fenlake.

lanu Green Lane, Forty Foot Lane.

leac-tun Leighton Buzzard.

leah (a) Appley, Apsley End, Aspley Guise, Crawley, Crawley Green, Farley, Lee, Leys, Oakley, Priestley, Shenley Hill, Streatley, Thurleigh, Wharley End, *Whyperley*, (b) Chibley, Hatley, Pedley, Puddle Hill, Stopsley, (c) Early Grove, Nares Gladley, Keeley, Riseley, Runley, Steppingley.

mæd (a) Shortmead, Wootton Broadmead, (c) Bushmead.

mægð, mægden (n) Medbury.

mæl Maulden.

mearc (c) Warmark.

meox Mixeshill.
mersc Marston.
mor Birchmore, Moor.
myln South Mills.
næss (c) Haynes.
ofer (b) *Buckleshore.*
*peac (n) Pegsdon.
pearroc Park Farm.
pirige Perryhill.
pol (b) Cople.
pott (n) Potton.
*răc (n) Reach.
rið Tingrith.
riðig (b) Efferiddy.
sand (b) Chicksands.
scearp Sharpenhoe.
scylf Shelton (2).
sealh Salph End, Salford.
sele The Hasells.
sloh (n) The Slough, Slough Wood.
snæd (b) Whipsnade.
stapol (a) *Stapleford,* Staploe, (c) Dunstable.
stoc (a) Stoke Mill, (b) Redbornstoke (?).
stocc (a) Staughton, (b) Redbornstoke (?).
stodfald Stotfold.
stow (c) Elstow.
stræt Stratford, Stratton, Streatley, Warden Street.
sul(h) (n) Souldrop.
swan Santon.
þeod-weg Ede Way.
þing Tingrith.
þorn Thorn, Thickthorn.
þorp Souldrop, Thrup End.
þyrne (b) Chawston.
topt The Toft.
*tot-ærn (n) Totternhoe.
tree (b) *Wixamtree.*
*treowiht (n) Trevor.
tun (a) Beeston, Campton, Chalton (2), Clifton, Dunton, East End, Eaton Bray, Eaton Socon, Etonbury, Everton, Houghton Conquest, Houghton Regis, Kempston, Leighton Buzzard, Luton, Marston M., Milton (2), Newton, Potton, Santon, Shelton (2), Staughton, Stratton, Sutton, Westoning, Wootton, (b) Chellington, Clipstone, Wyboston, Wymington.
turf (n) Turvey.

wæd Biggleswade.
weald Harrold, Wold.
weard Wardhedges, Warden, Warden Hill.
wearg Warmark (?).
weg Frenchmans Way, Mixeshill, Wickey.
welig Willbury, *Willey*, Willington (?).
whippeltre ME, *Whyperley*.
wic (a) Astwick, *Dilwick*, Flitwick, Harrowick, Hinwick, *Holli-wick*, Wick End, Wickey, (b) Goodwick, *Kinwick*, (c) Tilwick.
wielle (a) Cauldwell, Chadwell End, Holwellbury, Ladywell, Radwell, (b) Bidwell, *Cogswell*, *Feaks Well*, Ickwell, Kimber-well, Sewell.
wince Wingfield (?).
worþ (a) Eyworth, (b) Barwythe, Colmworth, Colworth, Ed-worth, Kensworth, Tebworth, Tilsworth, Wrestlingworth.
*wræst (n) Wrest.
wrong (n), ME Gorerong.
wudu (a) Hay Wood, *Hazel Wood*, King's Wood, Lowe's Wood, Northwood, *North Wood*.

(b) Huntingdonshire

bæc, bece Ashbeach, Chalderbeach.
bearu *Barrow*.
beorc Brick Mere.
beorg (a) Weybridge, Barham, (c) Lattenbury.
bigging Biggin.
broc (a) Brook End, (b) Billing Brook, Tilbrook, (c) Conington Brook, Hinchingbrooke.
brycg (a) Farcet Bridge, Hail Bridge, (b) Botolph Bridge.
burh (a) Bury, Littlebury, (b) Alconbury, Eynesbury, Lans-bury Farm.
burna Morborne.
busc (n) (b) Warboys.
byge (n) (c) Milby.
ceaster (b) Godmanchester.
cot(e) Caldecote (3), *Caldecote* (2), Coton Barn, Cotton, Ley-court, Ugmere Court.
cros (b) Normancross.
dæl Humbrel's, Ruddles Lane (?).
(ge)delf King's Delph.
denu (a) Waresley Dean, (b) Agden, Buckden, Gransden.
dic (b) Wornditch.
dierne *Dernford*.
dræg *Draym..e*.

dun (b) Haddon, Huntingdon.
ea Wray.
ear (n) Earith.
eg (a) Goldiford, Northey, Rowey, Woolvey (?), (b) Bodsey,
Higney, Ramsey, (c) Horsey Hill.
ende Over End.
fæled (n) *Falt*.
fearr Farcet.
fenn Farcet Fen, New Fen, Stocking Fen, West Fen.
fleot Fletton.
ford (a) Coppingford, *Dernford*, *Goldiford*, Hartford, Offord,
Wansford, (b) Hemingford.
gata Stangate Hill.
geac (n) Yaxley.
glæd (n) Glatton.
golde(n) *Goldiford*, Goldpit (?).
graf Grafham.
grafa (a) Old Weston Grove, Stow Grove, (b) *Hepmangrove*,
Rooks Grove.
grene The Green (2), Green End (2), Green Farm.
grund Stanground.
(ge)hæg (a) Harthay, Woolvey (?), (b) Dudney, *Edwoldeshey*,
Gamsey, *Suershay*.
ham (a) Barham, Grafham, (b) Bluntisham, Wintringham,
(c) Somersham.
hamstede (a) *Beachampstead*.
heafod (a) Farcet.
healh (a) Parkhall.
heordewic Hardwick, Hardwicke, Hardwicks, Eynesbury Hardwicke, Monks Hardwick.
here Hartford.
hind *Hind Lake*.
hlaw Stirtloe.
hoh (a) Calpher, The How, Midloe, Rushey, Southoe, Ward
Mound, (b) *Baldewynho*, *Mulsoe*, *Ravenshoe*.
holm Bromholme, Holme, Port Holme.
holt *Rawerholt*.
hragra (n) *Rawerholt*.
hungor Hungry Hall Cottages.
hyll Hill Farm, Houghton Hill, Pidley Hill.
hyrne The Herne.
hyrst Old Hurst, Woodhurst.
hyð Earith.
ing Lymage.
ingas (b) Gidding, Yelling.

ingtun Alwalton, Brington, Conington (?), Covington, Diddington, Dillington, Ellington, Elton, Stibbington, Wennington.
kaupmaðr Coppingford.
lacu *Hind Lake*, Lake Brook.
(ge)lad (a) Crollode, Fenton Lode, Yaxley Lode, Monks' Lode,
 (b) Hook's Lode.
land Kingsland, Redland.
leactun Leighton Bromswold.
leah (a) *Ockley*, Prestley, Sapley, Stonely, Stukeley, Woolley,
 Yaxley, (b) Abbotsley, Aversley, Pidley, Poplar, Waresley,
 (c) Raveley, Washingley.
lundr (a) Holland, (b) Toseland.
mere (a) Brick Mere, *Draymere*, Ramsey Mere, Trundle Mere,
 (b) Ugg Mere, Whittlesey Mere.
mor Middle Moor, Staughton Moor.
næss Outerness.
ofer Orton.
ord *Sword Point*.
pearroc Parkhall, Paxton (?).
pirige East Perry, West Perry.
pol *Wallpool*.
port Port Holme.
*ripp (n) Ripton.
rið (a) Sawtry, (b) Meagre.
sæppe Sapley.
scealfor (n) Chalderbeach.
sealh Salome Wood.
sealtere (n) Sawtry.
slæd Dipslade.
slæp Slepe.
stan (a) Hurstingstone, Leightonstone, (b) *Ogerston*, Keyston.
steort Stirtloe.
stigel Stilton.
stocking Stocking Fen.
stow Long Stow, Wistow.
strod Stroud Hill.
styfic Stukeley.
sweora *Sword Point*.
þorn Thorn.
þorp (a) Upthorpe, (b) *Sibthorpe*.
þyrne (a) Bythorn.
trendel (n) Trundle Mere.
tun (a) Brampton, Broughton, Chesterton, Denton, Easton,
 Fenton, Fletton, Glatton, Hilton, Houghton, Leighton, Water
 Newton, Newtown, Orton, Ripton, Stanton, Staughton,

Stilton, Upton, Walton, Weston (3), *Woodhoughton*, Wyton,
(b) Boughton, Kimbolton, Sibson, (c) Conington, Hamerton,
Paxton, Woodstone, Wyton.

(ge)wæsc (a) Washingley (?), (b) *Arnewas*.

weald (a) Weald, Weybridge, Old Hurst, Old Weston, (b)
Bromswold.

weard Ward Mound.

weg Broadway.

wic (a) Wigan, Wistow, (b) *Brightamwick*, Winwick, (c) Spaldwick, Worlick.

wielle Broadall's (?), Holywell, *Wallpool*.

wielm (n) Wansford.

worþ (b) *Blasworth*, Buckworth, Catworth, Folksworth, Molesworth, Needingworth, Tetworth.

wudu Monks' Wood, Upwood, Wistow Wood.

NOTES ON THE DISTRIBUTION OF
THESE ELEMENTS

A few notes upon the distribution of certain p.n. elements
may be given, but as the comparative material for other counties
is not (except in one or two cases) complete as yet, the remarks
can to some extent be only tentative.

botl. In passing east from Bucks and Northants we have
passed right away from the use of this element. No example has
been noted in either county.

burh. *Bury* in the manorial sense is rare in Hunts but is
fairly common in Beds where it tends to take the place of the
type of manorial name with final possessive *s* which is so common
in Ess, Herts and Bucks. The contrast is that between *Grimsbury* and *Gastlings*.

burna as the name of a small stream is a good deal less frequent than broc in these counties.

camp. Beds can now be added to the list of counties in EPN
14 (*s.v.*) in which this element is found.

cot(e). Intensive study of these counties puts them along
with Berks and Northants as counties in which there is a high
proportion of such names. In Hunts this is due especially to
the frequency of the name *Caldecote*.

ende. The great frequency of the word 'End' is a marked
feature of Beds nomenclature on the modern map, and early
forms enable us to carry back this way of denoting the outlying
districts in a parish to the 13th cent.

grene. It may be noted that the substantival use of *Green*

can, on the Beds and Hunts evidence, be carried back to the
13th cent., much earlier than is suggested in EPN or in the
NED.

ham. In Beds Newnham, Biddenham, Bromham, Clapham,
Pavenham, Felmersham are on the Ouse, Blunham on the Ivel,
while Studham and Higham Gobion are on high ground away
from any river. In Hunts Bluntisham is on the Ouse, but the
other -*hams* are not on streams of any importance. All except
Newnham (Beds) and Wintringham (Hu) are recorded in DB.

hamstede. That places which had names ending in this element were not of great importance is indicated by the fact that
both of the places with it have been lost in Hunts and one in
Beds, while two out of three were lost in Bucks.

hamtun. Of this suffix, of which one example was found in
Bucks, no trace has been found in Beds or Hunts.

healh is almost unknown in Hunts and very rare in Beds.

heordewic is specially common in these counties, berewic is
unknown.

hoh. Beds is distinctively the county in which this suffix is
most common. Though it is considerably smaller in area than
Bucks, nearly four times as many examples have been noted,
while Hunts, which is half the size of Bucks, has just as many
examples of this suffix as that county. In the frequent use of
this element Beds and, to a less extent, Hunts go along with
Northants. It is difficult on grounds of topographical difference
to suggest why this element should be thus frequent in these
areas.

ing and ingas are not common and it is noteworthy in Beds
(except for Knotting) to what tiny places the names refer.
Knotting, Gidding and Yelling are alone recorded in DB.

ingaham. Wintringham (Hu) is the only example in the two
counties.

ingtun. There is a group of these names—Cardington, Goldington, Stevington and Willington—just to the east of Bedford
itself. The only other example in the county is Podington in the
north-west. In Hunts the distribution is not so clearly marked.
Alwalton, Elton and Stibbington are on the Nen, Brington and
Ellington on a tributary of Alconbury Brook. The others are
Diddington, Dillington and Wennington. There are two
examples of *ingatun*, Seddington on the Ivel and Lidlington on
high ground near Ampthill. All except Seddington (Beds) and
Wennington (Hu) are on record in DB.

leah is fairly common in Beds, almost as common as in Bucks.
It is quite common in certain areas of Hunts, but naturally is
rare in the fen-districts.

mor. It is noteworthy how few examples of this suffix there are in either county.

stoc. We are practically out of the *stoke*-area in these counties.

þorp is very rare in both counties. In Beds it is certainly of English origin and probably is so in Hunts, though we cannot be sure.

tun. The proportion of names with this element is much the same as in the counties bordering on them. Apart from the names in ingtun it is noteworthy how few of the place-names in tun are compounded with a pers. name—only five out of thirty-two in Hunts and four out of thirty-one in Beds.

worþ. In the frequency of this suffix Beds and Hunts differ markedly from Northants, still more so from Bucks. They agree closely on the other hand with Cambs and Herts. In all cases except one it is compounded with a pers. name. All except Barwythe and Colworth (Beds) and Needingworth and Tetworth (Hunts) are on record by the date of DB.

PERSONAL-NAMES COMPOUNDED IN BEDFORD-SHIRE AND HUNTINGDONSHIRE PLACE-NAMES

Names not found in independent use are marked with a single asterisk if their existence can be inferred from evidence other than that of the particular p.n. in question. Such names may be regarded as hardly less certain than those which have no asterisk. Those for which no such evidence can be found are marked with a double star.

(a) *Bedfordshire*

Ælf(a)	*Elvedon*
Ælfrēd	Efferiddy
Ælfrīc	Arlesey
Ælfsige	Halsey
**Ælne, **Ællen	Elstow (?)
Bacca	Backnoe
*Badel	Battlesden
Bæra, Bære	Barwythe
Bassa	Basmead
Bēda	Bedford
*Biccel	Biggleswade
Billa	Billington
Biscop	Biscot (?)

*Blæcc(a)	Bletsoe
**Bluwa	Blunham
Bola, Bolla	Bolnhurst, *Bolnoe*
Brūna	Bromham
Bucca	*Bucklow*
*Buccel	*Buckleshore*
Bud(d)a	Budna
Byda	Biddenham, Bidwell
Cada	*Cadbury*, Caddington
*Cǣg(a)	Keysoe, Cainhoe
*Cærda	Cardington (?)
*Cealf	Chawston
Cēn	Kensworth
Cēnrēd	Cardington (?)
Ceobba	Chibley
Cēolwynn (f)	Chellington
*Cicca	Chicksands
*Cnotta	Knotting
*Cocc(a)	Cogswell (?), Cople (?)
*Cogga	Cople (?)
*Col	Colesden
Cola	Colworth
*Cuca	Kitchen (?)
*Culma	Colmworth
Cyneburh (f)	Kimberwell
Cynemund	*Kinwick*
*Cȳta	Kidney (?)
Deule (ME)	Duloe (?)
Dudewine	Dedmansey
Ealda	Arnoe
Earna	Early Grove (?)
Ed(d)a	Edworth
Eopp(a)	Hipsey
*Fæcc	Feaks Well
Feolumǣr, Fealamǣr	Felmersham
**Fōt	*Fotsey*
Fráni (ON)	Francroft
**Gāl	Galsey (?)
*Gicca	Ickwell
*Glǣda	Gladley (?)
*God	Gadsey Brook (?)

Gōdgifu (f)	Goodwick
Golda	Goldington
**Hætta	Hatley
Hagena	Haynes (?)
*Hān	Hanscombe
Herela	Harlington
Hocga	Hockliffe
Hræfn	Ravensden
Hrani (ON)	Renhold (?)
Hrōc	Roxton
*Hutta	Upbury (?)
Klyppr (ON)	Clipstone
Lēof	Lewsey
*Lihtla	Leagrave (?)
Līnbeald, Lindbeald	Limbersey
Madalperht (OHG)	Meppersall
*Papa	Pavenham
Passa	Pascombe
*Pearta	Pertenhall
*Peol, *Piol	Pillinge
Pic(el)	Picts Hill
*Polla, Pulla	Polehanger (?)
*Pott	Potsgrove
Puda	Puddle Hill
**Pudda, Poda	Podington
*Pulloc	Pulloxhill
Putta	Putnoe
Pydda	Pedley
Rǣdburh (f)	*Redburnstoke*
*Rīsa	Riseley (?), Risinghoe (?)
Scīr(a)	*Shirdon*
*Scyttel	Shitlington
Secga	Segenhoe
*Seofa	Sewell
*Sifel	Silsoe
Stakkr (ON)	Stagsden (?)
Stēapa	Steppingley (?)
*Stopp	Stopsley
*Styfa	Stevington
*Sunna	Sundon

**Teobba	Tebworth
*Þýfel	Tilsworth
Tudda	Toddington
Wada	*Wadlow*
*Wændel	Wensdon
*Wæra	Warmark (?)
*Wibba	Whipsnade
Wīdmund	Wymington
Wīgbeald	Wyboston
Wīhstan	*Wixamtree*
Wil(l)(a)	Wilshamstead, Wilden, Willington (?)
Wint(a)	Wingfield (?)
Wintra	Winter Wood
*Wrǣstel	Wrestlingworth
*Wrocc	Roxhill

(b) Huntingdonshire

Acca	Agden
Æþel	Elton
Æþelbeorht	Aversley
Æþelweald	Alwalton
Bealdwine	*Baldewynho*
Bicca	*Beachampstead*
Billa	Billing Brook
**Blǣge	*Blasworth* (?)
*Blunt	Bluntisham
Bōtwulf	Botolph Bridge
Brūn	Bromswold
Brȳni	Brington
Bucc	Buckworth
Bucge (f)	Buckden
Buga	Boughton
Byrhthelm	*Brightamwick*
*Catt	Catworth
Cofa	Covington
Cun(n)a	Conington (?)
Cynebeald	Kimbolton
Duda	Dudney
Dudda	Diddington
*Dylla	Dillington

Ēadweald	*Edwoldeshey*
Ealdbeald	Abbotsley
Ealhmund	Alconbury
Ēanwulf	Eynesbury
*Earna	*Arnewas*
Eli	Ellington
Folc	Folksworth
Garbod (OGer)	Gamsey (?)
*Gārmōd	Gamsey (?)
**Giella	Yelling
Godmund	Godmanchester (?)
*Grant(a)	Gransden
Gūðmund	Godmanchester (?)
*Gyd(e)la	Gidding
Hædda	Haddon
*Hǣðmund	*Hepmangrove* (?)
Hēahmund	*Hepmangrove* (?)
Hemma, Hemmi	Hemingford
*Hnydda	Needingworth
Hōc	Hook's Lode
Hræfn (OE), Hrafn (ON)	Ramsey, *Ravenshoe*
Hrōc	Rooks Grove
Hunta	Huntingdon
*Hycga	Higney
*Hyncel	Hinchingbrooke
Ketill (ON)	Keyston
Klakkr (ON)	Clack Barn
Mæðelgār	Meagre
Mūl	Molesworth, *Mulsoe*
Ordgār	*Ogerston*
*Perruc	Paxton (?)
**Poppa	Poplar (?)
Pyd(d)a	Pidley
Sibba (OE), Sibbi (OE, ON)	Sibson, *Sibthorpe*
Sigeweard	*Suershay*
*Stybba	Stibbington
**Sumor	Somersham (?)
*Sunmǣr	Somersham (?)

*Tetta	Tetworth
Til(l)a	Tilbrook
Toglos, Tóli (OScand)	Toseland
Ubba	Ugg Mere
*Wǽr	Waresley
*Wassa	Washingley (?)
Wearda	Warboys
*Wenna	Wennington
Wina	Winwick
Wintra	Wintringham
Witel	Whittlesey Mere
**Wud	Woodstone
Wurma, *Wyrma	Wornditch (?)

FEUDAL NAMES

(a) *Bedfordshire.* Aspley Guise, Eaton Bray, Cockayne Hatley, Higham Gobion, Houghton Conquest, Houghton Regis, Leighton Buzzard (?), Marston Moretaine, Milton Bryant, Milton Ernest, Westoning.

(b) *Huntingdonshire.* Abbot's and King's Ripton, Orton Longueville and Waterville, Offord Cluny and Darcy, Sawtry St Judith.

MANORIAL NAMES

(a) *Bedfordshire.*

(i) *bury*-names: Arlesey and Aspley Bury, Etonbury, Grimsbury, Grovebury, Holwellbury, Howbury (?), Mossbury, Mowsbury.

(ii) *Possessive*-forms: Bowels Wood, Beckerings Park, Corbetshill, The Creakers, Fernels Wood, Gastlings, Redding's Wood, Scroup's Farm, Someries, Thralesend, Traylesfields, Wake's End, White's Wood, Zouche's Farm.

(iii) Forms without the possessive *s*: Birchfield, Daintry and Exeter Wood, Ducksworth, Middlesex Farm, Pippin and Temple Wood.

(b) *Huntingdonshire.*

(i) *bury*-names: Lansbury.

(ii) *Possessive*-forms: Bevill's Wood, Gaynes Hall, Moynes Hall.

(iii) Forms without the possessive *s*: Daintree Farm, Bulby Hill.

(iv) Pseudo-manorial: Humbrel's Farm.

FIELD AND OTHER MINOR NAMES

In collecting material for the interpretation of the place-names (i.e. those found on the O.S. maps) a good deal of material has been gathered in the form of field and other minor names, especially those of boundary marks. It is impossible to deal with these exhaustively, first because they are too numerous, and secondly because many of them are without much interest, consisting largely of forms which are common in all field-names; further, it is but rarely that one has a succession of forms in an individual name such as is usually necessary if any satisfactory interpretation is to be attempted. A selection alone can be attempted. Unfortunately also the material is not at all evenly distributed. Manors belonging to Dunstable and Newnham in Beds can be studied in great detail and so also can the Ramsey manors in Hunts and Beds, not only in the Ramsey Cartulary but also in the extensive series of early Court Rolls which have survived. Other manors are much less well represented in early documents and for a good many of them no material at all has been discovered.

An analysis of these elements, with illustrations of their use, follows. Those elements that have already been fully illustrated in the true place-names are for the most part left unnoticed.

æcer is very common in both counties. Such a name as *Horsacre* (2) suggests that it was not always used of arable land. Among other compounds we may note *Stubbedhalfacre* (1371), *Goldenehalfaker* (13th).

OFr anglet (n), 'little angle,' is found in *les Anglettes* (1219), *les Aungletes* (1307) in Weybridge Forest.

ME balke is common, as in *Rowebalk* (1254) in Sharnbrook. It is used of a ridge of turf between two cultivated strips.

banke is found occasionally in Hunts, never in Beds.

bekkr. One example has been noted in Clophill (Beds), *Walebek* (1273), and one in Eynesbury (Hu), *Holebek* (1279).

bigging. *Neubigging* by Tempsford (1329).

ME brade is fairly common in Beds, as in *Le Redebrade* (1406) in Houghton Regis, *le Prestysbrade* (1426) in Holme, *Stratebrade* (c. 1330) in Stondon, *Russebrade, Merebrade, le Goresbrade* in Haynes (1309). It presumably describes a particularly broad cultivated strip. Seeing that it can appear as late as 1607 as *the Brade* in Gravenhurst, it must go back to OE *brǽdu,* 'breadth,' rather than the adj. brad. That on the other hand must lie behind *le Brode* (1392) in Kempston. *v.* Addenda.

bræc, brec in the form *brache* or *breche* is common in Beds and Hunts.

broc is very common both in Beds and Hunts, e.g. *Buterwellebroc* (1225), *Maniwellesbroc* (1252).

ME broile, a French word, ultimately of Teutonic origin, and frequently found in old hunting districts, is found once in Weybridge Forest (1219), cf. Broyle (Sx). It was used of 'a park,' 'a warren stored with deer.'

burna is used but rarely in either county as a stream-name element.

ME butte is common and is used of a short strip or set of short strips, of uncertain length, ploughed in the angle where two furlongs meet at right angles.

camp is occasionally found, but only one example has been noted in Hunts.

ceart seems once to be found as *Churt* in Cardington (13th cent.).

croft is exceedingly common in both counties. It is often compounded with the owner or holder's name. At other times the first element indicates the crop as in *Riecroft, Lincroft, Garscroft, Watecroft* (from hwæte), *Benecroft, Haycroft, Madecroft* (from mæd), *Henepcroft* (OE *henep,* 'hemp'), *Barlycroft,* or the animals belonging to it as in *Calvescroft, Piggescroft, Fesauntescroft,* or the soil, as in *Grenecroft, Smethecroft, Stonecroft. Caumpecroft,* which must contain camp, is interesting. It apparently denotes a croft, or enclosure, occurring within a stretch of open field.

cumb has been noted thrice in Beds.

dæl has been noted in a few names in *-dale* in both counties.

dal is fairly common as *dole* in Beds and very common in Hunts. *Cotmannedole* in Sharnbrook (13th) is worthy of note as containing OE *cotmanna* (gen. pl.), 'cottagers.'

dam in *Pideledam* (1280) and *Jarwelledam* (1301) can thus be carried back farther than in the NED.

dell is occasionally found in Beds, as in *Rutheresdelle* (13th), *Brixisdelle* (1225).

denu is common in both counties.

Y

draf, ME *drove*, is found in Hunts, as in *le Overdrove* in Brampton (c. 1307). It is still used in the fens for a road along which cattle are driven, and for a water-way.

feld is clearly used, at least until the 14th cent., entirely of open unenclosed country, for it is nearly always compounded with *North, South, East* or *West* or with *New* or *Middle*. It is not until the 14th cent. that we get it compounded with a pers. name as in *Ylgerisfeld* (1326).

ME ferye. *Blauncheferye* in Fletton in 1279 seems to refer to a white 'ferry' of some kind across the Nen and to carry that word back some 150 years earlier than it is recorded in the NED.

gara is very common in field-names.

(ge)hæg, ME *heye*, is very common in old woodland or forest areas.

haga is rare. In Hunts it may be the ON *hagi*.

ham is occasionally found in both counties but is never compounded with a pers. name, e.g. *Brocham, Foxham, Stonyham, Hecham, Bradenham*.

hamm has been noted three times in Beds and once in Hunts.

hamstede is found in *Chalfhamstede* in Eaton Socon (1359).

OE heolstor (n), 'darkness, hiding-place, retreat,' is found as the first element in *Hulsterdene* (1387, Cl), *Husterdoune hole* (1602, BHRS viii. 57) in Northill. *v.* Addenda.

hlaða seems to be found once in Hunts in *Kaldemowelath* (1351).

hlinc is found in both counties.

holmr, ME *holme*, is fairly common in both counties but specially so in Hunts.

horn is occasionally found in both counties.

hrycg is very rare in both counties.

hryding has been noted in a few cases in Beds.

hulc, 'hut, hovel,' is found in two examples of *Hulkestede* in Hunts, and in *Hulkeyerd* (1294) in the same county.

ME hullok in *le Hullokes* (1309) in Haynes carries the word *hillock* back nearly a century earlier than in the NED.

hyll rather than hlaw is the common word for 'hill' in both counties. We have early examples of *Hungerhills* in Flitwick and Biggleswade in Beds and Ellington in Hunts.

læs is occasionally found in Beds, very rarely in Hunts. It is compounded with *Ox, Horse, Cow* and *Summer*.

land is one of the commonest elements in field-names in both counties. The first element may describe (*a*) the shape of the 'land' or strip, e.g. *Wowelond, Wrongelond,* 'crooked, twisted,' *Holewelond, Longelonde, Schortelond, Endleslondes, Thortelond,* probably '*thwart*-land,' (*b*) what grows on it, e.g. *Henepland,* 'hemp-land,' *Linland,* 'flaxland,' *Peselond, Barlilond, Watelond,* 'wheat-land,' *Berelande,* 'barley-land,' *Benelond, Ruschelond, Bexlond,* 'box-land,' *Flexland, Rielond, Thornylond, Brereslond,* (*c*) the condition of the soil or the slope of the ground, e.g. *Blakmoldiland, Stonilond, Rouland,* 'rough land,' *Strodland,* 'marshy land,' *Hangindelond, Gorilond,* 'muddy land,' *Eredlond,* 'ploughed land,' *Morland,* 'swampy land,' *Hangrelonde, Waterlond,* (*d*) the occupier or lack of such (pers. names are comparatively rare), e.g. *Nomanneslond, Almuslond,* 'alms-land,' *Chircheslond, Shereveslond, Pottereslond, Akermanlond, Hydemanneland, Fremanneslond.*

mæd is very common in both counties.

pi(g)htel is occasionally found in both counties.

plot (n) is twice found in Hunts, the names being *Inlandeplottes* (1252) and *Madplot* (1300) from mæd.

pytt. We have references to *Lampittes*, from which 'loam' must have been taken, *Chiselput,* 'gravel pit,' *Sandputtes, Turpettes*, probably 'turf-pits,' and to more than one *Wolfpit*, presumably snares for these animals, e.g. *Wlpetes* in Dillington (1248).

rand is very rare, the only example that has been noted is *Lullesfordehiderande* in Beeston (Hy 3).

riðig is very common in bounds, etc. in the form *ridye, riddy,* etc., but hardly ever survives.

rod first appears in Hunts, some five examples having been noted.

sceat, later *shot,* so common in field-names in other parts of the country, has only been noted once, in Hunts.

seað has been noted in Beds, in Stagsden, Toddington and Whipsnade.

sic has been noted once in Beds as *Bradesike* (1302) and once in Hunts (1227) as *Stonhyll Syke*.

slæd is fairly common in both counties. Among the compounds are *Maydenslade, Watereslade, Chalkesslade, Russeslade, Depeslade, Welleslade, Hayslade, Groveslade.*

*snoc (n). This rare p.n. element, discussed in *Essays and Studies by Members of the English Association*, iv. 67, is found in 1279 in *Ravelesnok* in Wood Walton, aptly describing the sharp point or *snook* made by the boundary of the two parishes in question just north of Lodge Farm.

stede. This element, apart from its use in *hamstede*, does not seem to occur in Beds field-names but is found several times in Hunts where it seems specially to refer to the actual site of some buildings, what in Scotland would be called a *steading*. Thus we have two examples of *Cotestede* (*v.* cote), a *Tunstede*, and two examples of *Hulkestede* from OE *hulc*, 'hovel.' There are also compounds *Gangsted*, *Spychsted*, whose meaning is less clear, and a *Ryngstede*, meaning presumably 'circular site.'

stocking is fairly common in both counties.

stræt is fairly common for roads or tracks which certainly can never have been Roman roads of any kind. We have a *Leuestrete* in Dunstable, a *Bradestrete* in Henlow, a *Heystrate* in Northill, Wrestlingworth, Old Weston, *Bothildestrate* in Ravensden, *Northamestrate* in Ridgmont, *Smalstrate* in Sharn-brook, *Cutbedestrete* in Totternhoe, *Brunestrete* in Haynes, *Haringstrate* in Stilton. Is this last a road along which *herrings* were brought and is it a name for Ermine Street at this point?

stycce first appears in Hunts, where some seven examples in the form *steche*, *stych* have been noted, e.g. *Horstych*, *le horne-destiche*, *Silkenestych*.

topt, toft is only found twice in field-names in Beds (in Sharnbrook and Stotfold), but has been noted in Elton, Grafham, Houghton and Warboys, compounded in one case with the Anglo-Scandinavian pers. name *Osebern*.

vangr. This Scand. element first appears in Hunts, except for one example in Beds in Sharnbrook, and is fairly common in the form *wang* or (more commonly) *wong*.

vrá is found once in *le Wro* in Milton Bryant (1247).

weg is the common term for a track, and we have reference to (*a*) the surface in *Clayweye, Grenewey, Stanywey, Stratweye*, (*b*) its users or whither it led in *Kilneweye, Fenwey, Chyrcheweye, Portewey, Woldweye, Milneweye, Fordeweye, Colyereswey, Chapelweie, Carteriswey, Wudewey, Thefwey, Lordesway, Dossereswey*, i.e. apparently one used by animals carrying a pack or panier, (*c*) the general character of the road in *Holewey, Biggewey, Riggeweye*, (*d*) what was carried along it in *Flexwey, Riscweg*, (*e*) its being on a boundary in *Mareweye. v.* Addenda.

wic is fairly common in both counties. Only very rarely is it compounded with a pers. name.

wudu. *Loswude* in Warden (Hy 2) is an interesting compound with hlose.

Among miscellaneous names we may note *Chrystetrewesdole* (13th), which must have taken its name from a 'Christ-tree' or crucifix, and a solitary *Hauenebyryelis* (sic) in Hemingford in 1300, a relic of some old heathen Anglian cemetery. There are two examples of the enigmatic *Kattesbreyn* noted in PN Bk 127 (both in Beds but on opposite sides of the county). We have a somewhat Biblical *Seuenebrethren* in 1217 in Wyton, a *Marchileslane* in Cranfield in 1300, evidently a lane leading to the practising ground for archery, *marchils* coming from OE *miercels*, 'mark to shoot at.' Other names are *Doggetail* (13th), *Penybarecoles* (1584), *Undep Lane* (13th), *Wilderwash* (1607), clearly a low-lying meadow where the 'wilder' or wild animals watered, *Gosebath* (13th). In *Thrumlonge* and *Ouerethrum* we seem to have an adaptation of *thrum*, 'piece of waste thread or yarn,' including the unwoven ends of the warp. We also have three independent field-names—*le Snape* (1301), *le heegh Snappe* (1331) and *Little Snap* (1227)—which may be the ME *snape*, discussed by Ekwall (PN La 17), a Scandinavian loan-word of uncertain origin and meaning, but the form with double *p* may be of different origin and be derived from a different stem. If it is related to the verb *snap*, 16th cent. *snappe*, and means 'small piece, fragment,' the history of that word is carried back much farther than the references in the NED would suggest.

PERSONAL-NAMES IN FIELD AND OTHER MINOR NAMES

(a) *Old English*

Ælfgȳð	*Aluithebrigg* (1202)
Ælfnōð	*Alnothescroft* (1217)
Ælfrēd	*Alfredeswelle, crucem Aluredi* (13th)
Ælfstān	*Alstoneslake* (1279)
Ælfweald	*Alfwoldesmede, Alfwoldeswey* (13th)
Æðelbeorht	*Aybrichesdale* (1252), *Aylbriktest* (1248)
Æðelflæd (f)	*Alfledestoft* (13th)
Æðelgifu (f)	*Ailivemade* (1252)
Æðelmǣr	*Ailmaresheie* (13th), *Aylmerewik* (1320), *Almeresgore* (1295)

Æðelnōð	*Elnothescrofte* (13th)
Æðelwīg	*Alewieshou, Aylwyslad* (13th)
Æðelwine	*Alwineshei, Alwynescroft* (13th), *Alwynesleye, Ayllenecroft* (1252)
Bacca	*Bakenhey* (1202)
Bacga	*Bagenhale* (1252)
Bada	*Badewrth* (13th)
*Badela	*Badeligford* (1248)
Bassa	*Basselawe* (1240), *Bassecroft* (1251)
Bēaggeat	*Beiatteslawe* (1217)
*Bearda	*Berdestapel* (1260), *Berdeley* (1313)
Beorhtsige	*Brixwyk* (1351), *Brixisdelle* (1225)
Beorhtstān	*Bricstaneswlle* (c. 1350)
Beornheard	*Bernardescroft* (1252), *Bernardesholm* (1300)
Beornwine	*Berewynesdene* (13th)
*Bibb(a)	*Bybeswell* (1318)
Blæcman	*Blecchemaneshul* (13th)
*Boden	*Bodenesdych* (1300)
Bosa	*Bosemere* (13th)
Bōthild (f)	*Bothildestrete* (13th)
Botta	*Bottenhale* (1307)
**Brant	*Branteswyrð* (937), *Brantesdone* (1251)
Brūn	*Brunescroft* (13th)
Cēolhelm	*Chelmescote* (1227)
*Col	*Colesac* (13th)
*Dægel	*Deilesford* (Hy 3)
Denegȳð (f)	*Denegiðegraf* (10th)
*Dēorling	*Derlingeshauedlond* (1300)
**Dinni (?)	*Dinneshangra* (937)
Dodd	*Doddesworthe* (c. 1350)
Dodda, Dudda	*Doddelowe* (1351)
Dudemann	*Dudemannesmere* (1252)
Dynne	*Dinneshangra* (10th)
Ēadburh (f)	*Adburnewell* (1286)
Ēadgifu (f)	*Edgyuecroft* (13th)
Ēadmǣr	*Admereshey* (1254)
Ēadrīc	*Edricheslenge* (1252)
Ēadweald	*Edwoldeshowe* (1276)
Ēadwine	*Edwinesmede* (1279), *Edwinescroft* (13th)
Ealdmōd	*Eldemodescroft* (1259)
Ealdrēd	*Haldredesbuttes* (13th)
Ealdrīc	*Aldricheshegge* (13th)
Ealdwine	*Ealdwining baruue* (11th)
Ealhfrið	*Ealferðeshlæw* (937)
Godric	*Godricheshulle* (13th)

Godwīg	*Godewystockynham* (13th)
Gūðmund	*Guthmundescrochet* (13th)
Hæddi	*Haddeswurth* (1262)
· Hagen(a)	*Hagenescroft* (13th)
Hicci	*Hykkesfeld* (1351)
Hild	*Hildismer* (1389)
Hildegār	*Hildegaresdike* (1219)
Hōc	*Hokescroft, Hokeseth* (13th), *Hokislond* (c. 1230)
*Hodd(a)	*Hoddecroft* (1309), *Hoddysdole* (1416), *Hodewyk* (1366)
Hudd	*Huddescroft* (1348)
*Hund	*Hundeslowe* (13th)
*Hūnhere	*Hunerestocking* (12th)
Hūnstān	*Hunestanesdich* (1262)
*Hutt	*Huttesbutt* (1219)
Hygebeald	*Hibaldesle* (1227)
Lēofic	*Leuechesford* (13th)
Lēofmann	*Lefmannesyate* (13th)
Lēofrūn (f)	*Leverunhey* (1235)
Lēofweald	*Lewoldeswelle* (13th)
Lēofwine	*Leofwinesgar* (1012), *Lefwinbery* (13th), *Lewyneshegdole* (1301)
Lifing	*Lfingeshauedlond* (13th), *Levyngeshege* (1392)
Lihtweald	*Litwoldusmere* (1406)
*Locc	*Lokkesgroue* (13th)
*Luddoc	*Luddokesmere* (1202)
Lull	*Lullesford* (Hy 3)
*Macca	*Maccanho* (1012)
Mūl	*Moulescroft* (13th)
Mussa	*Musseridi* (13th)
Passa	*Passelowe* (1392)
P(e)ada	*Padeworth* (1351)
*Prætt	*Pretteslane* (c. 1350)
Pyttel	*Pytelewworth* (1366), *Putlesho* (1247)
Rūmbeald	*Rumboldesbecc* (1012)
Sǣmǣr	*Semareswonge* (1252)
Sǣmann	*Semaneshauedlond* (1318)
Sǣweald	*Sewaldescroft* (13th)
Scot	*Schotteshauedlond* (1337)
*Seofa	*Seuerod* (13th)
Sigebeorht	*Syberdeshul* (13th)
Sigelāc	*Silokesmed* (1219)
Sigenōð	*Sinodeslak* (13th)
Sigewine	*Sywinesdole* (1226)

*Sīðhūn	*Sithuneshegge* (13th)
*Tetta	*Tetlowe* (1406)
*Wændel	*Wendyllesbroc* (1463)
Wealhwine	*Walwynneslond* (13th)
Wīgmund	*Wymundeswelle* (1259), *Wymundeswong* (13th), *Wimundeshul* (13th)
Wīg-ræd	*Wirdeshale* (13th)
Wihthere	*Wytteresgore* (13th)
*Wilgeat	*Willietescroft* (1252)
Willoc	*Wylokescroft* (1252)
Wulfhere	*Wolveresheye* (1366)
Wulfmær	*Wullemarescroft* (13th)
Wulfnōð	*Wolnodeswey* (1209), *Wolnotheslenge* (1252), *Wlnodescroft* (Hy 3)
Wulfrīc	*Wlfricheswelle* (1202)
Wulfstān	*Wolstonescroft* (1252), *Ulfstonedikes* (1348)
Wulfweard	*Wulwardescroft* (1252)

(b) Scandinavian

Ásbjörn	*Osebernestoft* (1313)
Ásketill	*crucem Anketilli* (13th)
Ásmundr	*Asmundemere* (1279)
Gunnhildr (f)	*Gunneldiscroft* (13th)
Gunrið (OSw)	*Gondridescroft* (1272)
Haraldr	*Haraldeshegges* (13th)
Íre	*Yresmere* (13th)
Káti	*Catieshege* (1252)
Ketill	*Ketelescroft* (1217), *Ketelesbrug* (13th)
Kippa	*Kyppescroft* (1392)
Klakkr	*Clackesmor* (13th)
Sveinn	*Swayneshul* (E 3), *Sueineswelle* (Hy 2)
Svertingr	*Swerthingcroft* (1222)
Þórketill	*Thurkillislond* (1219)
Þýri (f)	*Thyrethorn* (1292)
Tóki	*Tokieshegge* (13th)
Tóli	*Tholeshey* (1230)
Wigot (ODan)	*Wygodeswey* (13th)
Ǫgmundr	*crucem Agmundi* (13th)

(c) Continental

Durand	*Durandescroft* (13th)
Everard	*Everadesfeld* (1367)
Fulk	*Fulkescroft* (13th)

Gerard	*Gerardesholm* (1244)
Gilbert	*Gilbertesdolle* (1240)
Hunfrid	*le Hunfriesgore* (1307)
Ilger	*Ylgerisfeld* (1326)
Lambert	*Lamberdesmer* (1284)
Otelin	*Otelynescroft* (13th)
Rainald	*Reynodlescroft* (1397)
Roger	*Rogereslond* (13th), *Rogeresholm* (1299)

INDEX

OF PLACE-NAMES IN BEDFORDSHIRE

The primary reference to a place is marked by the use of clarendon-type.

INDEX

OF PLACE-NAMES IN HUNTINGDONSHIRE

z

INDEX

OF PLACE-NAMES IN COUNTIES OTHER THAN BEDFORDSHIRE AND HUNTINGDONSHIRE

[References to Buckinghamshire place-names are not included as these have been fully dealt with in the volume already issued upon the names of that county.]